Radio Control with 2.4 GHz

GW01311291

Die Deutsche Nationalbibliothek v
der Deutschen Nationalbibliografie, ... bibliografische
Daten sind im Internet über dnb.d-nb.de abrufbar.

Impressum:

Roland Büchi, 2014

Herstellung und Verlag: BoD - Books on Demand, Norderstedt

ISBN: 978-3-73229-340-7

German edition:
'2.4-GHz-Fernsteuerungen', publisher: vth, Verlag für Technik und
Handwerk neue Medien GmbH

Preamble

In the first years of the new millennium, in the development of radio control technology with 2.4 GHz, a real revolution has taken place. While the frequencies in the MHz range were standard over many years, a large proportion of modern communication technology and with it radio control (RC) has since taken the band between 2.4 GHz and 2.48 GHz. This development was possible not least due to faster transistors and today's microprocessor technology. The biggest advantage for the user is that he no longer needs to coordinate with others if he wants to switch on his radio control. All transmitters transmit on the same frequency band, and the receivers know through a sophisticated technology which signals are dedicated for them.

Model construction has been equipped with RC for over fifty years. Without it, this fascinating hobby would probably never have inspired generations of people. Without RC, especially the airplane, helicopter, car, and ship models would not have achieved such popularity. And this book is aimed primarily at the users of these types of models.

Of course there were many developments, especially with RC on the MHz frequency bands, which led to the technology used now. Since today almost exclusively 2.4 GHz RCs are sold, this technology is concentrated on in this book.

The components of a modern radio control, such as the transmitter, receiver, servos, and other peripherals, are discussed in a separate chapter. It is also shown here how the various units can be combined into a complete and functioning system.

This leads immediately to the range and its properties. On its own, this is not yet meaningful, because different factors such as obstacles, reflections, or the movement of the model itself can interfere with the safe transmission at 2.4 GHz. In addition, different antenna types with different characteristics are used. Their optimal alignment is also discussed on the basis of many examples. Many 2.4 GHz RCs operate with diversity, i.e. with multiple antennas. It is also specifically discussed here how the antenna orientations can be optimized in this mode. Of course some comparisons with the signal transmission in the MHz range are always made.

There is also a chapter on the modulation and transmission types. In many data sheets of modern RC terms such as PPM, PCM, ASK, FSK, PSK, FHSS, FASST, DMSS, or DSSS are found. The book is designed to help so that by the end the reader can classify and understand these properly and knows their most important characteristics.

The user interfaces of today's RC are mostly two sticks. Each of them can be moved in all four directions. The stick allocations to the functions are clearly defined in the various models. However, there are several allocation modes. These are discussed, and a look into the future will also be made, with new concepts of user interfaces which may be capable of replacing the sticks. Rotary encoders, switches, and buttons are also discussed.

The data transmission is possible in both directions with the 2.4 GHz technology. This means that the receiver in the model can also operate as a transmitter and can transmit important telemetry data from the model during operation. In doing so, the transmitter functions briefly as the receiver and may optionally display battery voltages, engine temperature, or currents of the model. An overview of the available systems is given.

At the end of the book are some practical tips on installing the system into the model, and for the interference suppression of components.

Contents

1. Introduction

The development of technical model construction has always been very closely linked with the development of radio controls. With 'actual' cars, airplanes, or boats, the driver, pilot, or captain sits inside. He controls them directly via sticks, steering wheels, brakes, accelerator pedals, throttles, etc. Since models in contrast are unmanned, they must be controlled from the outside. Since the radio control represents the only method of intervention to the models in use, it has become the most important aspect. Contemporary models of all types are getting larger and larger, and ever more powerful motors are being installed. The largest models sometimes cost as much as a midsize car. Their weight and speed also represent a significant safety risk. Only the reliability and capabilities of modern radio control systems allow safe maneuvers on land, at sea, and in the air.

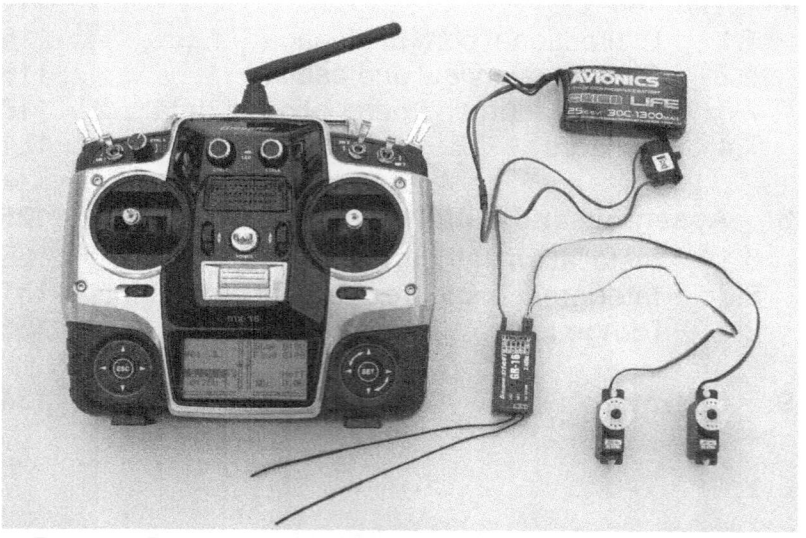

Figure 1: Radio control system with all necessary components

Today, at the beginning of the 21st century, almost all commonly used RCs are based on 2.4 GHz technology. This is actually only a logical development of all previous transmission types. It is even so that the origin of many features of modern RC is several decades old.

1.1 Origins of wireless transmission

The birth of wireless transmission was approximately in 1864. At that time, James Clerk Maxwell predicted the existence of electromagnetic waves in theory. In 1886 the physicist Heinrich Hertz was able to prove this with measurements. Guglielmo Marconi could use them practically in 1901. He was able to send and receive wireless Morse code across the Atlantic Ocean. Electromagnetic waves can propagate at the speed of light in vacuum and in the air. They can transport energy and also information from a transmitter to a receiver unit, without any wires.

First points of contact to model construction
As with many other technologies, model construction was one of the first applications here. Nikola Tesla, the genius in electrical engineering, made many inventions in various fields. At the world exhibition in New York in 1898, he presented a radio-controlled model boat. That was probably the birth of the RC in model construction. But another 50 years passed before RC became more common. The first transmitters, which sent in the MHz range, were sold in the 1950s.

1.2 From the MHz to GHz transmission

The range of MHz transmitters is in the order of several hundred meters to a few kilometers. It depends on the permitted transmission power. If two transmitters want to transmit their signals within the same area using the same frequency range, then the receiver can't distinguish where the detected signal originates from. The received signals are then an overlapping of the two transmitter signals and it is not able to detect the proper signal. The regulatory authorities therefore had to legislate for the use of frequencies. RC in model construction thus received its own frequencies in the MHz range. They are the only users of these.

The path to GHz radio controls

For many years from the 1960s to the 2000s, frequency panels dominated in model construction events. Before the model pilot or captain could switch on his radio control, he first had to mark his channel on the panel so that no one else would use it. Any other pilots with the same channel could not switch on their radio control during this time, not even for testing purposes. Here, the discipline of everybody was vital. Often transmitters were still switched on unintentionally. This was usually very bad, especially in model flight. More than a few models were brought rather rudely from the sky in this way.

There were also cheap RCs, which were very broadband due to technical imperfections. Thus they sent their signals on the neighboring channels in addition to their own. These could be identified only with special frequency scanners. The scanners were indispensable tools of the organizers, especially for larger events.

First 2.4 GHz RC

The development of microprocessors also influenced transmission technology. In modern communication technology, the band of 2.400 GHz to 2.483 GHz is very important. It belongs to the so-called ISM (Industrial, Scientific, Medical) band. RC became an additional application of it. At these 2.4 GHz frequencies, however, the RC is not alone. Unlike the MHz band at which a certain frequency range is reserved exclusively for them, the 2.4 GHz RCs must share this band with a variety of other modern means of communication. A popular generic term for this is WLAN, short for Wireless Local Area Network.

In the early 2000s, many 2.4 GHz conversion kits were sold for the MHz RC. Here, only the high-frequency part of the transmitter and the receiver had been replaced. The transmitter housing and peripherals such as servos and motor controller were still the same as before. However, it soon became apparent that the new technology had a decisive advantage compared to the old one: one was now no longer dependent on others not using the same frequency at the same time. One could therefore switch on one's radio control at any time regardless of other users. For this reason, the 2.4 GHz RC replaced the old MHz RC almost completely within just a few years.

Biggest advantage of the 2.4 GHz technology

The many means of communication in the 2.4 GHz band transmit their data only reasonably free from interference when the transmitter and the receiver use the same encoding. On the other hand, they also need to change the frequencies at any time and know from each other which frequency is being used at the moment. This is just a simple summary of what is discussed in more detail in Chapter 5.

This coordination between the transmitter and the receiver is called binding. Essentially, it is what replaces the insertion of the quartz pairs in the transmitter and receiver in the MHz RC. Somewhat loosely, it can also be explained as: the binding is used to set the language between transmitter and receiver. Only if both understand the language can they speak to each other. If many people are in a room and all want to communicate with each other, then it works if the Frenchman understands only the French sentences and the Englishman understands only the English ones. It works even better if all speak at the same volume. So technically, no one should speak with more power than the others. This too will be discussed in Chapter 5.

In any case, this has the consequence on the airfield, the track, or the lake that one transmitter speaks French with its receiver while another one speaks English. In transmission technology there are an infinite number of these 'languages'. And so it is (almost) impossible that in the same place two receivers and transmitters can't understand each other because a lot of others are communicating with each other on the same frequency band.

2. Components of a radio control

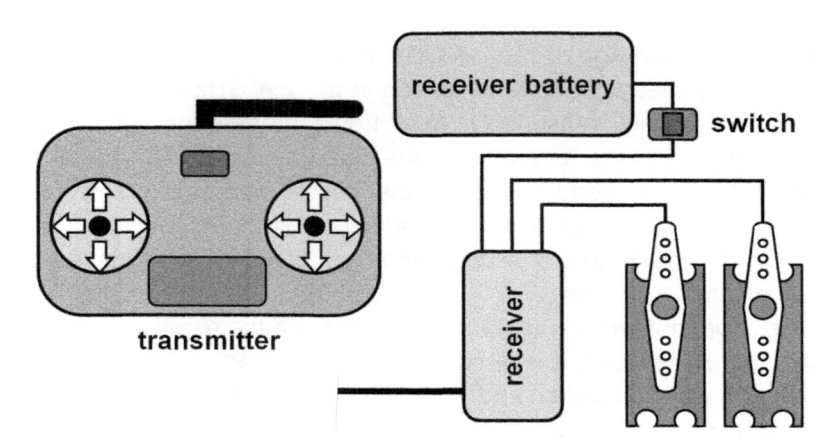

Figure 2: Schematic illustration of an entire RC system

An entire RC system consists of at least the components of Figure 1 in Chapter 1. In Figure 2, they are shown schematically. They are explained briefly here and will then be presented in more detail in the following subsections. On the left side the transmitter can be seen, and on the right side is all that is placed in the model. However, there are different ways to provide the receiver with energy. The version shown here is used for example in models with a combustion engine. Here, the receiver is powered directly with its own battery. Even if there is no motor, as with sailplanes, such a battery is required. However, the models are often driven by an electric motor. Then, the receiver is usually powered via the motor controller, directly from the drive or flight battery. So a separate receiver battery is not necessary. Such a configuration is shown schematically in Figure 3.

The servo is an indispensable element of the radio control technology. It ensures that the signals coming from the radio control cause corresponding actions in the model.

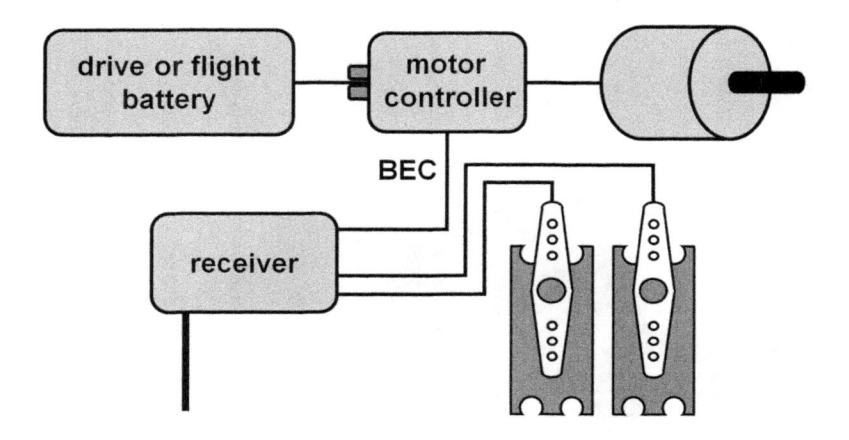

Figure 3: Schematic diagram of a receiver system, which is powered by the drive or flight battery

2.1 Transmitters

Handheld transmitter

The transmitter shown in Figure 4 is a so-called handheld transmitter. This is certainly the most common among all transmitter types. As the name suggests, the transmitter is held with the hands on both sides. Often the pilot controls only with his two thumbs, which are placed on the tops of the sticks.

Another method of control is to take the sticks between your thumbs and index fingers and clamp the transmitter with the palms. The author prefers this control mode, as it allows in his opinion a fine control. A strap that is worn around the neck can provide additional support. This offers the advantage that the transmitter cannot slip out of your hand and fall to the ground. But it has the disadvantage that it is sometimes cumbersome, for example during a throw start of a model airplane.

Figure 4: Handheld transmitter

Console transmitter

Mostly one controls with the thumb and index finger, while the hand is braced on the transmitter. The transmitter remains in position without the help of hands. For many handheld transmitters, extensions for console transmitters are available.

Whether a handheld or a console transmitter is preferred depends on the personal preferences of the pilot. Often it is the case that those who only need the four functions of the cross sticks prefer the handheld transmitter. If you need to control more functions, you have to operate different encoders or switches in addition to the two sticks. This is rather difficult with a handheld transmitter. Therefore, these pilots prefer console transmitters. Examples are large model airplanes or model ships, in which a lot of additional functions are often found.

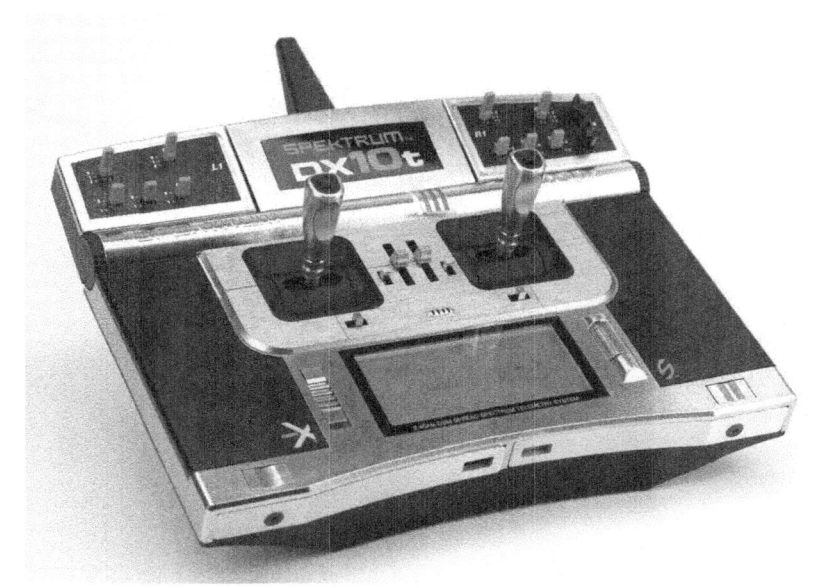

Figure 5: Console transmitter

Colt transmitter

A very different type of control has become popular with RC cars and fast RC boats. This is the colt transmitter of Figure 6. A right-handed person takes the transmitter in the left hand. With his right hand he operates the steering wheel. The push button can be operated on both sides and is 'gas' while pushing, and upon retraction 'brake' or 'backwards'. Both together can be implemented in the modern brushless motor controllers with one function or channel. Frequently, there is a third channel available. This is designed for mechanical brakes. With a sometimes additional fourth channel, the front and rear axles can be braked separately. Thus, RC cars can be controlled more close to reality than with cross sticks. However, the number of functions is limited; more than four channels are rare with Colt transmitters.

Figure 6: Colt transmitter

2.2 Receiver

The receiver (Figure 7) is the centerpiece on the model side. It takes the information via its antenna, processes it, and passes it to the peripherals such as servos and motor controllers. In the receiver there is a dedicated slot for each channel for these devices. The slots all have three pins. One is for the ground, one is provided for the power supply, which is approximately between 4.8V and 7.4V, and one is for the transmitted signal. As will be discussed in Chapter 5, with the 2.4 GHz technology there are different variants of data transmission from the transmitter to the receiver. However, between the receiver and the peripherals a pulse width modulated (PWM) signal has been transmitted for many years. Approximately every 20 ms a pulse with a length between 1.0 ms and 2.0 ms is transmitted to the servos and the motor controller. All manufacturers support this standard and therefore the peripheral devices are also replaceable. Modern microprocessor technology also allows additional functions. Thus, in some receivers 'fail safe' functions may also be programmed. If the receiver does not receive a meaningful signal due to interferences over a certain time, it can take a previously defined position with selected or even all servos. The modern digital servos can also take a 'fail safe' function by themselves. This will be described below. The receivers often also support a 'hold function'. Then they ensure in the event of a weak signal or a fault that the servos hold the previously taken position.

Figure 7: 2.4 GHz receiver

Often we speak today of so-called high-current-capable receivers. In most systems, the energy of the servos is distributed via the three-pin plug. Also in the largest models, this is performed in this way. Some of today's servos need currents of several amperes.

The supply line to the receiver must then provide the sum of the currents of all connected servos. This places high demands on the connector system. High-current-capable receivers then provide correspondingly thicker cable and connector systems which are designed for these high currents.

2.3   Servos

The servos are the real workhorses of an RC. They move proportionally to the stick position and transmit their position mechanically to the rudders, the wheels, the brakes, or to all that needs to be moved in the model. Figure 8 shows servos which are installed in a car model. The name is derived from the Latin. 'Servus' translates as 'slave'. The servo needs to do exactly what it is told by the receiver via the PWM signal. With its horn, it has to move to the position corresponding to the signal. Therefore, it always has an integrated position control. Originally this control was designed analogously. In modern servos it is also often performed digitally. This leads to the two types used today. Their properties are discussed in the next two subsections.

Figure 8: Servos installed in a model car

Analog servo

Figure 9 shows a block diagram of an analog servo. At the input the desired value is applied, which is the PWM signal discussed above. In the gearbox in the simplest case a potentiometer is used for position measurement. It detects the position of the rudder horn. The comparator is illustrated in the diagram with the +/−. Here, the PWM signal is converted into DC voltage with the use of filter technology. In technical terms, this is a low-pass filter. Depending on which of the two voltages is larger, the amplifier lets the motor rotate in the one or the other direction. When both the voltages are equal, the desired position is reached and the motor stops rotating. This kind of control and thus the servo are therefore called analogous, because the comparator and amplifier do not work with digital values '0' and '1', but with voltages between 0 V and 5 V (or up to 7.4 V). This system constantly adjusts to the position of the gearbox and thus the rudder horn. The PWM signal is only renewed every 20 ms, or about 50 times per second.

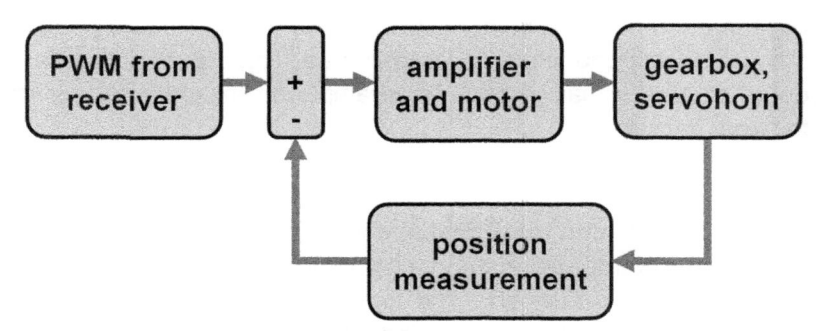

Figure 9: Block diagram of an analog servo

Digital servo

Basically, Figure 9 also applies for modern digital servos. However, the electronic system is controlled by a microprocessor. Optical sensors or Hall sensors are used in many versions as position sensors. Their signals are digitally read into the comparator. This also calculates directly the pulse length of the PWM signal. Thus, other pulse lengths and PWM signals can also be evaluated.

17

There are from various manufacturers also faster rates than 20 ms for the updating of the servo position. In some systems, the PWM signals can even be repeated three to four times faster. Digital servos can then often be switched to faster rates. Thanks to the higher update rate of the desired value, it is then possible for example to approach the position with a higher torque. With some servos you can also adjust the maximum current. However, you can't fool physics. Due to the higher torques advertised in the catalogs, power consumption also increases. That should be heeded in the context of digital servos. Thus several manufacturers recommend the use of larger batteries with the use of digital servos. However the microprocessor technology also allows additional settings which would not have been possible with analog technology. Thus, with different systems, a 'fail safe' or a 'hold' function at the level of the servo can be programmed, even with different priority levels. Therefore, during a short failure of PWM only the latest position is kept, and in a longer failure can also be moved into a safe position. This may be the idle gas position, for example in a model with a combustion engine.

But the often heard statement that a digital servo is more precise than an analog one is not necessarily true. The accuracy depends on other factors than 'digital', for example on the used sensor or the gear backlash.

Weight, torque, and rotational speed

These three words define the main characteristics of a servo. The weight depends on the size, the torque, and the rotational speed and lies between about 5 g and 100 g. The torque is often expressed in the unit Ncm for Newton centimeters. It is approximately in the range between 5 Ncm and 200 Ncm. To represent these numbers more understandably, a little thought experiment can be performed. If a servo can achieve a torque of 10 Ncm, then you can attach a weight of 10/10 kg = 1 kg at a distance of 1 cm from the axis of rotation. The divider 10 is thereby obtained because of the force of gravity, a physical size that depends on the gravitation. At a distance of 2 cm, it would still be half, so 500 g and at a distance of 3 cm, it would still be 333 g. For an extremely strong servo, with a torque of 200 Ncm, 200/10 kg = 20 kg could be attached at a distance of 1 cm from the axis of rotation. Figure 10 shows this with the attached weight.

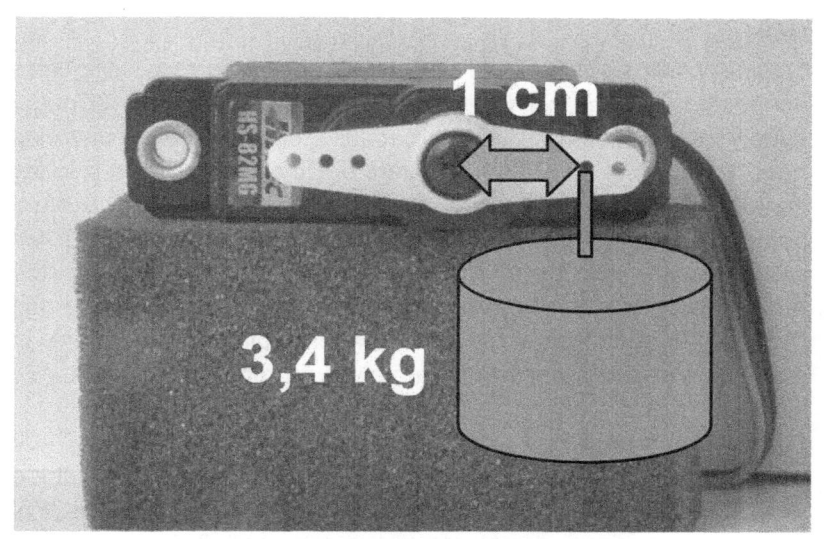

Figure 10: Servo with 34 Ncm torque

The servo shown in Figure 10 can supply a torque of 34 Ncm at a supply voltage of 6 V according to the data sheet. Thus a weight of 34/10 kg = 3.4 kg can be attached at a distance of 1 cm from the rotation axis. Some manufacturers differ in their data sheets between actuating torque and holding torque. Actuating torque is the torque that can be supplied by the servo in the rotating state. Holding torque is the torque at standstill. With this distinction, the holding torque is usually specified slightly higher than the actuating torque. The torque which the servo must supply is proportional to the power consumption. So a servo requires less power when it only needs to move a smooth rowing, as the needed torque is less then. So you should make sure that the rudder linkage can be moved well. For many applications, the actuator speed is important. This is correctly always specified in seconds per degree. If the travel is for example a quarter turn, or 90° or a total of +/− 45°, then it is often expressed in seconds per 45°. So that is the required time to move the horn from the center to one end position. This is approximately between 0.1 s and 0.2 s for commercially available servos.

Voltage
Originally the servo voltage was 4.8 V, because four cells were connected in series, each with 1.2 V. In Section 2.5 ('Energy supply') this is discussed further because today's receiver batteries have voltages of up to 7.4 V. Therefore the manufacturers today sometimes provide solutions at 4.8 V, 6 V, and 7.4 V, and sometimes also larger. The servos which can be operated at 7.4 V can of course also be operated at lower voltages. However, the other way around it often does not work. The possible torque and the rotating speed also varies at different voltages. Therefore, in some data sheets these values are given for different voltages.

Gearbox and bearings
In order to achieve a high torque, the servos are set with a gearbox. These are made of plastic in servos with low loads. For higher requirements, metal gearboxes are used. This should also be preferred when the variations of the torques are large or if the servos will be subjected to bumps and shocks. However, these gearboxes are slightly heavier and somewhat more expensive.
Even with the bearings of the shaft carrying the servo horn, there are different versions. With low requirements sliding bearings are used, for higher requirements the shafts have single or double ball bearings.

2.4 Motor controllers

As flying or driving motors, today brushless motors are almost exclusively used. A brushless motor needs a brushless controller because of the special control. At the input side, a brushless controller is supplied with a DC voltage by a flight or drive battery. It can be connected directly to the receiver like the servos. This provides it via a pulse width modulated signal with a speed value.

A pulse length of 1 ms corresponds to 'idle gas' and a pulse length of 2 ms corresponds to 'full throttle'.

The supply voltage is then converted into a three-phase AC voltage. When a motor should rotate faster, the frequency and the amplitude of the AC voltage have to be increased in such a way that the rotational speed is able to follow. So the motor controller does a variety of tasks. It must on the one hand evaluate the resulting PWM signal from the receiver and calculate the desired speed. On the other hand it has to convert the battery voltage into a three-phase AC signal and ramp up the motor speed to the desired speed.

The current development of the technology does not speak for the use of brushed motors in model construction. Their efficiency is much lower compared with the brushless technology and the brushes are in addition wear parts that will shorten the life of the motors. However brushed motors are easier to control because they do not require three-phase AC voltage, but only a DC voltage. For this reason, there are still some systems with brushed motors and corresponding motor controllers on the market, especially with cheaper RC cars or ship models.

Another very important additional task of the motor controller is the power supply of the receiver via the flight or drive battery. This is discussed in detail in the next section. Figure 11 shows a motor controller and a brushless motor. The three-pole plug is inserted into the receiver like with the servos. The receiver's energy supply is also routed through this connector.

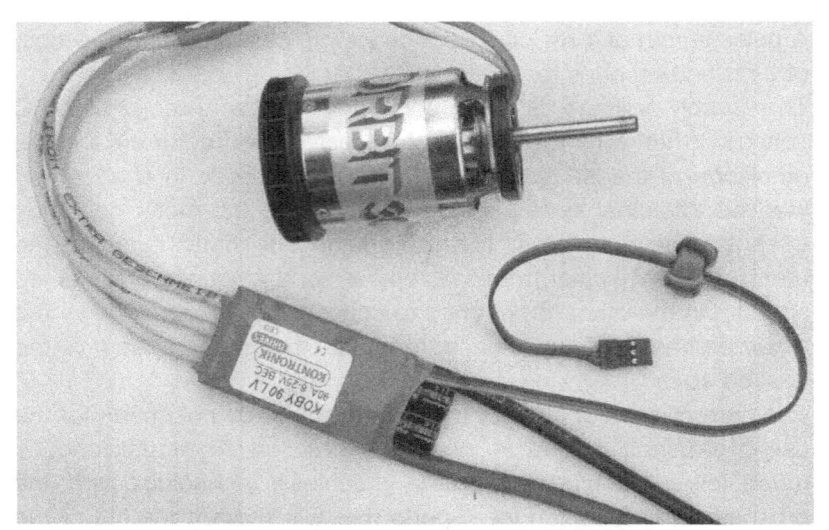

Figure 11: Motor controller and brushless motor

2.5 Energy supply

Transmitter battery

The energy supply of modern radio control systems is usually designed so that with a battery pack an operation of several hours is possible. Depending on the operating duration and types, in the transmitter, batteries are used with voltages between 5 V and 12 V and electric charges between 1500 mAh and 3000 mAh, and even more.

Since not every gram counts with the transmitter battery, the nickel-metal hydride (NiMH) battery is often still used. However this has some disadvantages. It has a lower energy density than lithium or LiFePO batteries, so it is heavier for the same energy. Furthermore, it discharges itself relatively quickly. It loses about 1% of its energy content per day. But there is also a big advantage. Its property with regard to insensitivity to deep discharge is however far superior to the modern types of batteries. A NiMH battery can sometimes last for years sitting in an unused radio control. It is quite amazing that it can then be recharged again and loses nearly nothing of its storage capacity.

Receiver supply with battery eliminator circuit

As discussed above, all models with an electric drive also have motor controllers. These are directly connected to the drive or flight battery. Thus it would be unnecessary for the receiver to get its own battery.

Therefore, it is now standard that all motor controllers have a so-called BEC, a battery eliminator circuit. As the name suggests, this is a circuit that eliminates the (receiver) battery. The object of the BEC is to produce a stable power supply for the receiver. To do so, the three-pin plug is plugged into the provided slot for the motor controller. One pole is the ground, one is the positive voltage supply, and one is the signal.

In the technical design of the BEC, there are two different types. One possibility is that a so-called linear regulator is used. It always cuts off at the required 5 V or 6 V. This causes a loss of power in the motor controller. It is calculated as the difference to the drive battery voltage multiplied by the current of the receiver and the servos. When for example an 11.1 V LiPo flight battery is used, and the receiver with its servos pulls a current of 2 A, then the power loss in the BEC is calculated as (11.1 V − 5 V) x 2 A = 12.2 W. One can easily imagine that the BEC in an RC car with a high torque and therefore high current steering servo generates a high power loss. On the other hand, when a battery with 40 V or even more is used in a large electric-powered aircraft, then the difference to the receiver voltage is so large that there is also high power dissipation.

For this reason in larger motor controllers, the BEC is often realized in the other variant, namely switched. It is then sometimes called SBEC in the data sheets, with 'S' standing for 'switched'. Such a BEC chops the battery voltage to the much smaller supply voltage for the receiver, using transistors. This happens with lower losses than in BECs with linear regulators. Thus, the heating of the motor controller is also limited.

The motor controller is also the energy center of the model. The secure energy supply of the receiver is then a more important task than the control of the motor. Namely, the controller must ensure that at low voltage the electric motor is switched off in time. So in all models and especially with airplanes the receiver must work all the time to ensure the secure control on landing.

Large models: separate power supply for the motor and receiver

The power supply via BEC actually only applies for small and medium-sized models. For large models, a separate drive or flight battery and a receiver battery are often used despite the added weight. This redundancy is created for safety reasons. As mentioned above, the power supply of the receiver is in most cases much more important than that of the motor. On the one hand, the technology with the BEC works very reliably today. On the other hand, the use of two separate power supply circuits for the motor and the receiver is nonetheless the safer option. For large models, the extra weight plays almost no role. This redundancy can be explained easily if the price of a large model comes in at the level of a midsize car. The loss of such a model because of receiver failure must be avoided at all costs. Also the safety risk would be high in such a case.

However, one has to disable the BEC voltage when using a separate battery. If the controller does not provide such an option, you have to disconnect the positive voltage from the BEC. Otherwise, the BEC voltage and the positive pole of the battery would be shorted via the receiver. Since the two voltages are never exactly the same, very large compensation currents would flow. These could damage both the receiver battery and the BEC circuit.

Receiver supply with battery

For models with combustion engines or without any motor, such as sailplanes or sailboats, no motor controller is available. In this case, a normal rechargeable battery is used as the power supply for the receiver. Again, the types of batteries NiMH, LiPo, or LiFePO are mainly used. As with the transmitter batteries, the NiMH battery is still often used at the beginning of the 21st century, because of its insensitivity to deep discharge. Indeed, there is no undervoltage shutdown in the receiver batteries, since the safe reception is always more important than the 'wellbeing' of the battery. In an emergency a deeply discharged rechargeable battery is far less serious than the loss of the whole model!

In the energy content, the range is greater than that of the transmitter batteries. There are also quite different models – very small or even large sailplanes or RC cars with combustion engines – which require a steering servo with high energy consumption. Batteries with voltages between 4.8 V and 7.4 V and electric charges between about 600 mAh and about 5000 mAh are available on the market. To calculate the maximum operating time, the following example is used.

Example 1: *A small sailplane has a receiver battery with a charge capacity of 1600 mAh. At the receiver and servos is flowing on average a current of 1.0 A. The maximum operating time should be calculated.*

The maximum operating time is calculated as 1600 mAh / 1.0 A = 1.6 Ah / 1.0 A = 1.6 h = 96 min.

Figure 12: LiPo and LiFePO, two modern receiver batteries

4.8 V to 7.4 V as supply voltage

Originally the receivers were supplied with four rechargeable NiMH cells, or four today no longer used NiCd (Nickel Cadmium) cells. Since these had a voltage of about 1.2 V each, the servo or other components were designed for 4 x 1.2 V = 4.8 V. Today, however, five NiMH cells or, as described above, other types of batteries are often used. The LiPo batteries have a voltage of about 3.7 V, which results in a voltage of 7.4 V with two cells in series. The LiFePO batteries have a slightly lower voltage per cell than the LiPo cells; two of them in series give about 6.6 V. Figure 12 shows the two energy sources.

The trend to slightly higher supply voltages is also a result of the increasing demands of the models, which are becoming bigger and bigger. The servos and other peripherals require more and more power. If one does not want to have too high a current, according to the physical law: power = voltage x current, this leads only to the demand of higher voltages. The older servos and motor controllers are generally compatible with today's RCs, as they're all driven by the PWM signal. However, whether they can work with higher voltages is not guaranteed. So if you want to be absolutely sure when using two cells LiPo or LiFePO, you should also make sure that the servos and motor controllers are designed for the higher voltage.

3. Properties of 2.4 GHz radio controls

In this chapter, first some basics of wireless data transmission will be considered. These are important for understanding Chapter 4 ('Antennas and their optimal orientation') and Chapter 5 ('Modulation and data transmission methods'). First, electromagnetic waves and their basic properties regarding the range will be considered. However, the range alone is not very meaningful, because in practice other things also affect the safe signal reception. Examples include the proximity of the antenna to the ground, obstacles, humid air, or the movement of the model itself. In order not to let the chapter appear too dry and theoretical, only very basic formulas are used. Moreover, they are always used with concrete examples with 2.4 GHz radio controls or also compared to the MHz radio controls.

3.1 Wireless data transmission using electromagnetic waves

An electromagnetic wave, which forms the basis of any wireless transmission of control commands, actually consists of two waves. As the name suggests, both an electric and a magnetic field are involved. An electric field has the unit 'volts per meter'. When such fields change their polarity with a frequency, as is the case here, then they are called alternating electrical fields. These cause alternating magnetic fields with the unit 'ampere per meter' and vice versa. This mutual interaction is in its entirety then the electromagnetic wave. Its signal propagation is often shown as in Figure 13. If it is sent at time $t_0 = 0$ of the antenna, it spreads out in a vacuum and in the air with the speed of light. After the time t_1, the wave front has a distance x_1. A little later, after t_2, it has that of x_2. The distance x of the wave front is generally calculated as

$$x = c \cdot t$$

where c is the speed of light, approximately 300,000 km/s.

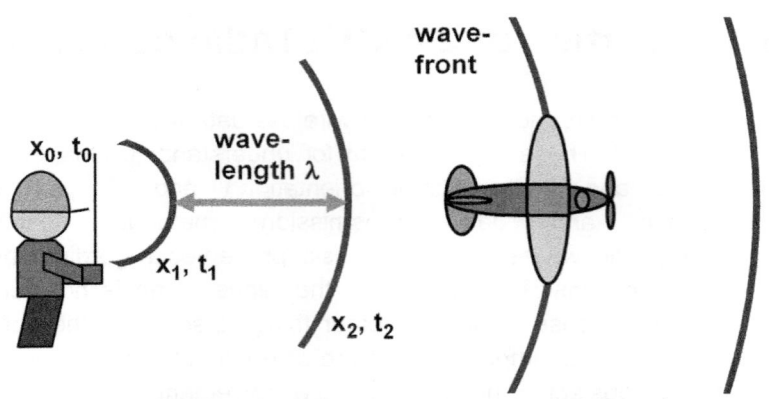

Figure 13: Signal propagation of electromagnetic waves

In order to generate the electromagnetic waves, a high-frequency generator is required. This generates on the antenna a sinusoidal alternating current at the desired frequency. A sinus has a minimum value and a maximum value. When the next maximum value of the wave is sent from the antenna, the last wave front is already a little further away. In the figure it is shown by the drawn lines. It raises the question how large the distance is between these wave fronts. The time interval of the maxima of a wave with a certain frequency is calculated as

$t = 1 / f$

Thus, using the formulas above, the distance between the wave fronts is

$x = \lambda = c / f$

This distance is called the wavelength and it is abbreviated by the Greek symbol λ (lambda). As a comparison, consider a stone that is thrown into water. Also there wave fronts are formed. They move in a circle away from the point where the stone entered the water. They become weaker and weaker and finally disappear altogether. Again, there is a distance between two wave fronts. A minimum between the two maxima can also be identified on closer inspection. Of course, the water wave propagates much more slowly, because it is not an electromagnetic wave.

Example 2

For radio controls with 72 MHz and 2.4 GHz, the wavelengths λ should be calculated. For 72 MHz a wavelength of (300,000,000 / 72,000,000) m = 4.17 m results. For 2.4 GHz there is a wavelength of (300,000,000 / 2,400,000,000) m = 0.125 m = 12.5 cm.

The light expands also with these waves at an almost incredible 300,000 km/s. This led to the hypothesis, which is now proved, that this is also an electromagnetic wave. Light has a much higher frequency and a wavelength which is in the visible range between 380 nm and 750 nm. One nanometer (nm) is one billionth of a meter, so 0.000000001 m. The comparison of GHz signal transmission with light is more important for the discussion below in Section 3.4. There the properties of the 2.4 GHz waves are explained, and the propagation of visible light can be explained in a way that is phenomenologically understandable to everyone.

3.2   Range and safe signal reception

The received signal level decreases with increasing distance. This in itself plausible statement is best explained again with Figure 13. The wave fronts of the propagating electromagnetic waves are illustrated here with segments of circles. Since this happens in three-dimensional space, there are actually spherical segments. The center of these is located at the antenna. In the vicinity of the transmitter antenna, the spherical segments are not so large – the length of the curve is meant here in the figure. With increasing distance from the center the curves are longer, since then the radius of the spherical segments is larger.

If the wave propagates in the air, then there is almost no loss of energy and in a vacuum there would be no loss of energy at all. Since each wave front transports the same energy, this is distributed with increasing distance from the transmitter to an ever-increasing sphere.

However, the receiving antenna is always of the same size, regardless of its distance from the transmitter. Thus, it can with increasing distance and greater spherical surface collect only a smaller and smaller proportion of the energy. The reception signal is therefore weaker at a larger distance. With a little geometry of the sphere, it is found that the reception level attenuates quadratically in proportion to the distance. This results in the following mnemonic:

If you double the distance from the transmitter, the level of the received signal is reduced to a quarter.

Once the receiver identifies an ever so small wanted signal, it can amplify it accordingly. However, there are various types of electromagnetic waves which exist in addition to the wanted signal emitted by the transmitter. On the one hand there are natural electric and magnetic fields, which originate from the earth, but also from the sun and galaxies. These are not of equal intensity at all frequencies, but certainly present in the 2.4 GHz range. On the other hand, there exist also man-made radiation sources. As discussed above, on the 2.4 GHz band the model pilot is indeed not alone, as there are also other users. This can for example be other radio controls.

Above a certain distance from the transmitter, the receiver can no longer detect any wanted signal, since this is too small compared to the noise. In technical language this is also called a too small 'signal-to-noise ratio'.

Signal fading with moving models

The range is calculated in the above mnemonic of free sight and without obstacles. It is specified in most common radio controls with at least about 1 km to 2 km. This is more than sufficient for the requirements in model construction. However, it is actually only of theoretical interest, and therefore not very meaningful by itself. In practice, other factors affect the safe reception. Therefore it is much more important that the reception within the practical distance between the transmitter and the receiver, in other words between the pilot and his model, is guaranteed everywhere and especially at any time, even under the consideration of reflections from the ground or water surfaces, or of obstacles. Both are very important for model construction. Car and ship models are always close to the ground or on the water and also model airplanes and helicopters are located near the ground at least during takeoff and landing as well as during a low-altitude flight.

On the other hand, obstacles such as trees or bushes for aircraft or a rough terrain for off-road models also belong to the normal conditions. In addition, the polarization described further below is constantly changing especially with airplanes, which can take all orientations, so the receiver antenna is always turning. However, the radio control must be working reliably in all of these cases.

Since the influence of all these factors changes continuously, especially with the moving models, the signal strength at the receiver is also changing. This effect is also called signal fading. The receiver must then adjust continuously the gain of the signal to evaluate it properly. As a result there are additional reception errors which affect the safe reception. Even a minimal reception signal must always be present, and taking into account all factors the worst signal must always be used as the measure. For the safe reception, there is another mnemonic:

For the safe reception, the range is only one of several factors. With moving models special attention must be paid to the signal fading.

Therefore, the following sections, 3.3, 3.4, and 3.5, are dedicated to the other factors for the safe reception and signal fading.

3.3 Fresnel zone

Basics

First, there is the question of how much 'air' or how much 'space' is even necessary to reliably transmit a signal from one point to another. Is there a line of sight needed, so the transmitter must be able to see the receiver, and if a line of sight is there, is it then sufficient if this exists only in a narrow range around the transmitter and receiver? These questions can be explained by the 'Fresnel zone' of Figure 14.

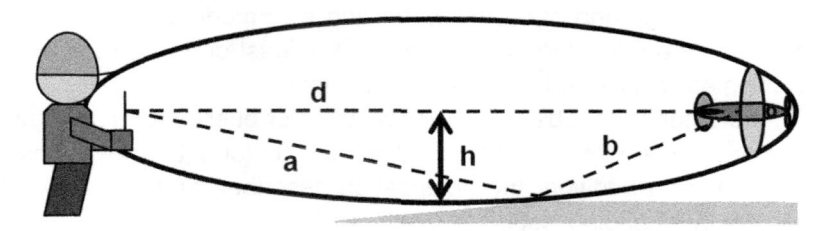

Figure 14: Fresnel zone

The Fresnel zone is an ellipse and its foci are the transmitter and receiver antenna. The radio control with rod antenna radiates with the vertical orientation as the figure shows, in the direction to the model with the maximum field strength, so along line 'd' (for distance). However, there is also a somewhat weaker emission in the direction of line 'a'. That is more specifically described in Section 4.2 ('Antenna radiation pattern and antenna gain'). Now it is thought that an object is located just on the edge of the Fresnel zone, such as the ground. When the electromagnetic wave 'a' occurs there, then it is reflected in proper topography right into the direction of the model. The wave which must cover distance 'a+b' has a longer path to the model than the wave which only has to cover distance 'd'. The Fresnel zone is now defined so that the following applies to each edge of the zone: $a + b - d = \lambda / 2$. This means that the wave reflected on the edge must cover half the wavelength more to the model than the wave which goes directly from the transmitter to the receiver.

However, then the wave front of the direct wave or the maximum arrives at the receiver together with the minimum of the reflected wave, so it will be weakened. The technical term for this effect is 'destructive interference'. Basically each wave which is reflected within the Fresnel zone causes a larger or smaller attenuation of the direct wave. Thus, the following mnemonic can be derived:

If the Fresnel zone is free of obstacles and also the ground is out of the zone, then there is safe reception of the radio control.

Now, however, immediately the question is raised of 'how wide' the Fresnel zone is. Using some geometric considerations a formula can be found for the maximum height h. According to the figure it occurs exactly in the middle between the transmitter and the receiver:

$$h = \sqrt{\frac{d \cdot \lambda}{4}}$$

The following small example will be calculated.

Example 3

For a 2.4 GHz radio control system the maximum height h of the Fresnel zone must be found for some typical distances between transmitter and receiver: 50 m, 100 m, 200 m, and 500 m. In addition a comparison with a 72 MHz radio control should be carried out for these distances. As the calculation of Example 2 shows, the wavelength λ = 12.5 cm for 2.4 GHz and λ = 4.17 m for 72 MHz. The above formula yields the following results for h:

	distance 50 m	distance 100 m	distance 200 m	distance 500 m
2.4 GHz	h = 1.25 m	h = 1.76 m	h = 2.5 m	h = 3.95 m
72 MHz	h = 7.22 m	h = 10.2 m	h = 14.4 m	h = 22.8 m

Table 1: Height h of the Fresnel zone for different distances at 2.4 GHz and 72 MHz

It is shown that the Fresnel zone is much smaller at the 2.4 GHz radio controls than is the case with the 72 MHz systems. Of course, the behavior at 27 MHz or 40 MHz is similar to 72 MHz. If you were to add even light as a comparison, which indeed is also an electromagnetic wave, then the height h of the Fresnel zone would only be a few millimeters for the distances in the table. As one can better imagine the behavior of light than that of the invisible 2.4 GHz waves, a little thought experiment will follow at this point.

In the darkness, a lamp with the distances from Table 1 is seen as a point of light. It doesn't matter if it takes advantage of all the free space for viewing, or if you look at the light through a thin tube which is only slightly larger than the light spot. This is only possible because the height h of the Fresnel zone is very small.

Fresnel zone in practical model construction

This raises the question of what this basically means for the practice of model construction. Figure 15 is intended to serve as an example for a model car. The conditions for ship models are similar; water also reflects the 2.4 GHz waves well, but the roughness of the water surface also leads to scattering effects.

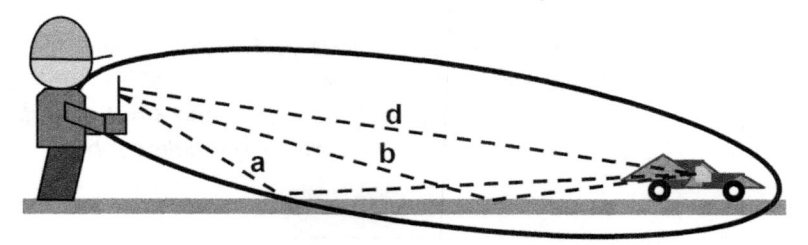

Figure 15: Fresnel zone for an RC car

In this case the waves a and b are both reflected within the Fresnel zone and contribute to the deterioration of the reception quality. To improve the reception signal, one must make sure that the ground extends as little as possible into the zone. The smaller the distance between the transmitter and a model is, the smaller is also h as shown in Table 1, and the less the ground extends into the Fresnel zone. Other benefits here are an elevated position of the model pilot and an elevated antenna mounting at the model.

Figure 16: RC car racing. The elevated standing position of the pilots not only helps the view, but also provides the model with better signal reception.

For model airplanes or helicopters it is often said in the manuals that care should be taken near the ground in respect to safe reception. Figure 17 shows this.

Again using the values of Table 1 for 2.4 GHz, it can basically be said that with airplanes – takeoff and landing aside – fewer obstacles extend into the Fresnel zone than with car and ship models. With a model helicopter which is located for example at a distance of 50 m from the transmitter, the height h is 1.25 m. This is approximately the height at which an average sized pilot holds the transmitter. If the helicopter flies at least this distance from the ground, the Fresnel zone is free of obstacles and the reception is well.

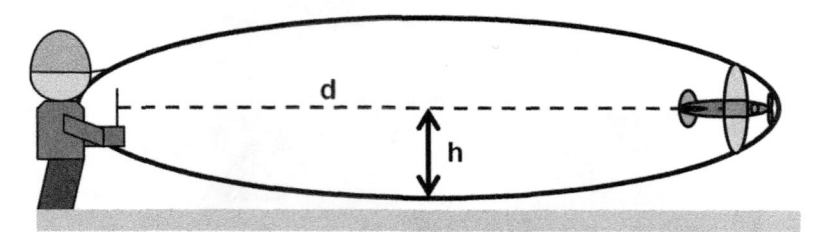

Figure 17: Fresnel zone for a model airplane

When the pilot of a model aircraft performs a low altitude flight at a distance of 100 m, the Fresnel zone is also free of obstacles if the distance to the ground is at least 1.76 m.

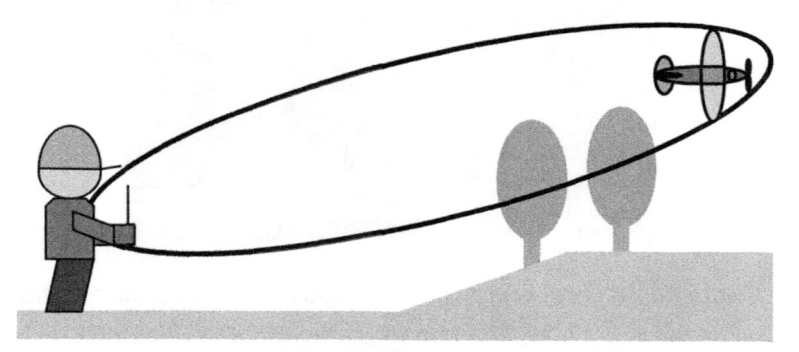

Figure 18: Bushes in the Fresnel zone

Figure 18 shows another case in practice with bushes in the topography. Generally speaking, that doesn't cause major problems if the view to the model is free and there is still some air between the bushes and the model. You may imagine that the Fresnel zone has only a height h of 2.5 m at a distance of 200 m. More critical, however, is the case where you are flying behind the bushes. In these cases there are obviously obstacles in the Fresnel zone. A critical case is also when a person passes close to the transmitter between pilot and model or even stops there. This then results in a radio shadow, which is described in the next section.

Sufficient reserve

In practice, however, it is not necessary that the Fresnel zone in each case is free of obstacles and that the topography must not extend into the zone. The transmitters operate at a high enough power that there are still some reserves. The above statements highlight that such obstacles or parts of the topography within the Fresnel zone can be a possible cause of poor reception or reception problems.

3.4 Radio shadow

When an electromagnetic wave hits an obstacle, a radio shadow or a dead spot follows behind. As described above, light waves are also electromagnetic waves and therefore behave similar to the 2.4 GHz waves. Once again, here the wavelengths are compared: 72 MHz results in a wavelength $\lambda = 4.17$ m, 2.4 GHz results in $\lambda = 12.5$ cm, and visible light is approximately $\lambda = 380$ nm to 750 nm, depending on color. For the visible light and its behavior any sighted person has extensive practical experience in everyday life. For example, everyone knows that there is a shadow when light falls onto an object.

This raises the question of what comparison could be done with the much longer waves of a MHz radio control. One possibility of comparison is the behavior of water. This is obviously not an electromagnetic wave. But it is also something which man can detect with his senses. If we imagine that water is flowing down a river and hits a rock as an obstacle, it fills in the gap immediately below this rock.

The behaviors of the invisible MHz and 2.4 GHz waves are located figuratively between water and light. However, one can argue that the MHz waves are more similar to water and the 2.4 GHz waves are more similar to light. What this means in practice for the model pilot is shown in figures 19 and 20.

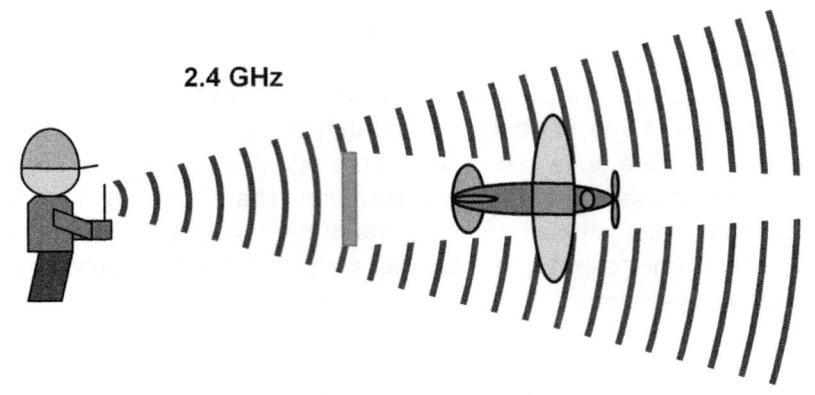

Figure 19: Radio shadow with a 2.4 GHz radio control

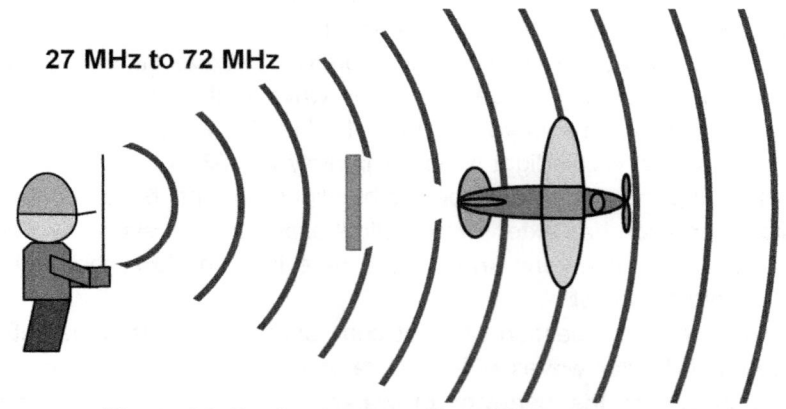

Figure 20: Radio shadow with a MHz radio control

The two figures are symbolic only, since in truth other effects occur (keyword diffraction) which are not discussed here. However, the general behavior can be demonstrated well. The illustration in Figure 19, which describes the 2.4 GHz waves, shows a relatively long radio shadow, precisely similar to that which occurs when light hits an obstacle. The illustration in Figure 20 describes the behavior of the MHz waves. Water as a comparison would close the gap again directly behind the obstacle. The MHz waves do this not quite directly, but the radio shadow is significantly smaller than that of the 2.4 GHz waves.

So in summary it can be said when comparing the 2.4 GHz to the MHz waves that the 2.4 GHz waves behave somewhat disadvantageously regarding radio shadow. However, with respect to the reflections near the ground, they have an advantage because the height of the Fresnel zone is smaller.

Rule of sight
Since the 2.4 GHz waves behave like light, this is a practical rule of thumb:

If there is a visual contact between the antenna of the model and the antenna of the transmitter, in the specified range reception is well in principle, since the GHz and the light waves are similar. Always be careful of obstacles between the transmitter and the receiver, in other words between the pilot and the model. As the pilot can't see the model behind the obstacles, so the receiver can only poorly see the signals of the transmitter. The result is a radio shadow.

3.5 Humidity, water, material penetration

Signal attenuation in humidity
Microwave ovens, which can be found in almost all kitchens, also emit electromagnetic waves in the GHz frequency range. Food consists mostly of water. The microwaves can only penetrate a few centimeters into the food before their energy is completely absorbed. This then leads to the desired warming.
What is useful in the above case of a warm meal is entirely undesirable in 2.4 GHz radio controls. On days with high humidity or rain, the tiny water droplets in the air can also absorb the GHz waves. However, this effect is small. It becomes more problematic if humidity forms a film of water on the antenna. This leads to significant signal attenuation.

Model submarines
There is therefore also a single segment of model construction in which the 2.4 GHz technology will never prevail despite its great

advantage over MHz radio controls of not having to coordinate any channels. With the radio control of submarines, the electromagnetic waves must penetrate the water. Waves in the GHz range can do this only over a distance of a few centimeters, so the wave is completely absorbed before it reaches the receiver. The submarine model builders must therefore ensure either that the upper part of the antenna is always located above the water, or that they stay with the MHz technology. The MHz waves are also strongly absorbed by water, but they can at least cover a distance of several meters up to the receiver.

Material penetration
Above, signal shadows and the 2.4 GHz waves were compared with light. This comparison has some limitations, especially with the penetration of solid materials, since light waves are simply much shorter than GHz waves. With models with closed fuselages without windows, there is no light at all. Therefore it is always dark in their interior. The longer 2.4 GHz waves also have their pitfalls here, but not quite as extreme as with light. A wooden hull in theory absorbs the 2.4 GHz waves more than the MHz waves. Also with the commonly used fiberglass hulls (fiber-reinforced plastic) the waves are more attenuated. In practice, however, the attenuation is manageable with these materials. The CFRP materials (carbon-fiber-reinforced plastic) or even carbon have to be carefully considered. These materials cause reception problems with the MHz waves already. However, this is quite problematic with the 2.4 GHz waves. The penetration in these materials is very poor. The correct positioning and installation of the antenna must be paid much more attention than with the long MHz antenna cables. Therefore with 2.4 GHz at least one antenna should always be placed on the outside of the model. In Chapter 8 ('Assembly and initial operation') some examples are discussed.

The length of the antenna is designed such that it ensures the best reception while it is freestanding. If it is in the vicinity of materials which attenuate the signals too much or if it is installed in the model in the vicinity of copper wires, it is also possible that it is electrically put out of tune. This then also has the effect that the received signals are weaker.

Findings in brief

At the end of the chapter the main findings are summarized again.

Electromagnetic waves are the basis of any wireless communication, also that of the 2.4 GHz radio controls. They propagate with the speed of light. If the distance from the transmitter is doubled, the level of the received signal is reduced to a quarter.

The operation distance is only one of several factors for the safe reception. Special attention must be paid to the signal fading of the moving models.

The Fresnel zone is an ellipse and its foci are the transmitter and receiver antenna. To prevent an attenuation of the received signal due to reflections, the Fresnel zone should be as free of obstacles as possible. Also the ground should not protrude too much into this zone.

The 2.4 GHz transmission is similar to the behavior of light. Behind an obstacle, there is a radio shadow. If there is a visual contact between the antenna of the model and antenna of the transmitter, the reception is well in the specified range.

Water absorbs 2.4 GHz waves more strongly than MHz waves. Also the penetration of solid materials is worse with 2.4 GHz. This should be taken into account especially when installing the antenna.

4. Antennas and their optimal orientation

In this chapter some general considerations about antennas will be made. These include the length and the polarization of electromagnetic waves. The radiation direction will also be discussed, and described with antenna patterns. Practical antennas for 2.4 GHz and their optimal orientation are discussed in Sections 4.3 and 4.4. In Chapter 3 it was calculated that the wavelength of the GHz radio control is much smaller than that of the MHz radio control. The GHz antennas are also generally much shorter than the MHz antennas. This property is based on the fact that for optimal function, the length of the antenna must be in the same order of magnitude as the wavelength.

Antennas are generally based on the so-called λ/2 dipole in Figure 21. In fact, here the high-frequency signal is fed in the middle of two λ/4 rods. The length of a λ/2 dipole for a 2.4 GHz radio control is equal to 12.5 cm / 2 = 6.25 cm, according to the calculation described in Example 2. Sometimes antennas work only with λ/4 (= 3.1 cm), so one half of the dipole is simply omitted. The other half is then the person holding the transmitter in their hand. These antennas are then also called monopole antennas.

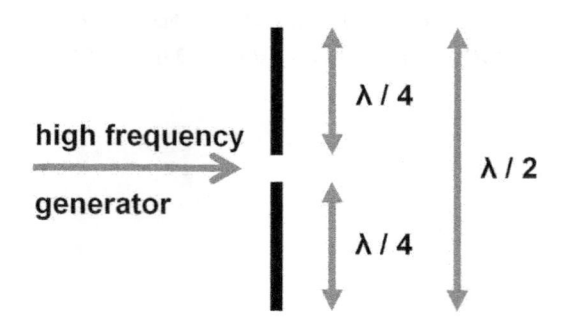

Figure 21: λ/2 dipole

4.1 Polarization

Electromagnetic waves are so-called polarized waves. This is very important for understanding how the transmitter and receiver antennas must be aligned. In Section 4.3, the meaningful antenna orientation is explained using practical examples. This is based on the polarization and the antenna radiation pattern, which is discussed in the next section. Figure 22 schematically shows the optimal alignment between a transmitter with a rod antenna and a receiver antenna.

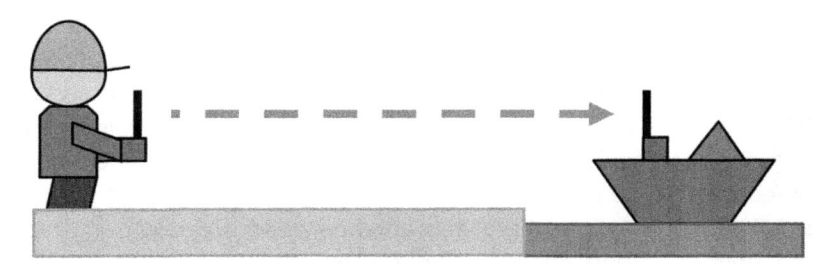

Figure 22: Vertical polarization of the transmitter and receiver antenna

The two antennas must therefore have the same orientation for optimal reception. Here the vertical polarization is shown. Thus the fact that the polarization direction always coincides with the orientation of the rod antenna will serve as a little mnemonic not to be confused with vertically and horizontally. The emission is at its maximum in the direction of the dashed arrow, i.e. in the horizontal direction in the figure. In the direction of the antenna itself, so in the direction of polarization, it is however minimal. In the optimum case, the receiver is also located somewhere at this height around the transmitter antenna. As a small exception, the receiver in Figure 23 should ideally not be behind the model pilot, since he causes a radio shadow with his body. That would worsen the reception signal. Due to polarization, the receiving antenna has the best possible reception if its orientation is vertically upwards or downwards.

4.2 Antenna radiation pattern and antenna gain

Antenna radiation pattern

The antenna radiation pattern is another important base for the Section 4.3. For its explanation the polarization can be assumed again as vertical. In the last chapter it is written that with a rod antenna with vertical polarization, the radiation is at its maximum in the horizontal direction. In the vertical direction, so in the direction of polarization, radiation is at a minimum. The exact distribution from maximum radiation up to no radiation is represented by the diagram in Figure 23. Another term for this is 'radiation pattern' or 'antenna beam'. The arrows show the emission at best. In the horizontal direction the arrows are the longest, thus the radiation is maximum there. If the model is located in this direction, it shows the best reception quality. It has to be mentioned that this is the antenna diagram of a λ/2 dipole. It is thus assumed that the signal is fed into the center of the antenna as shown in Figure 21. If only a λ/4 monopole were there, the antenna pattern would be flat on the bottom. If one again imagines the arrows in this case, they would be small towards the bottom.

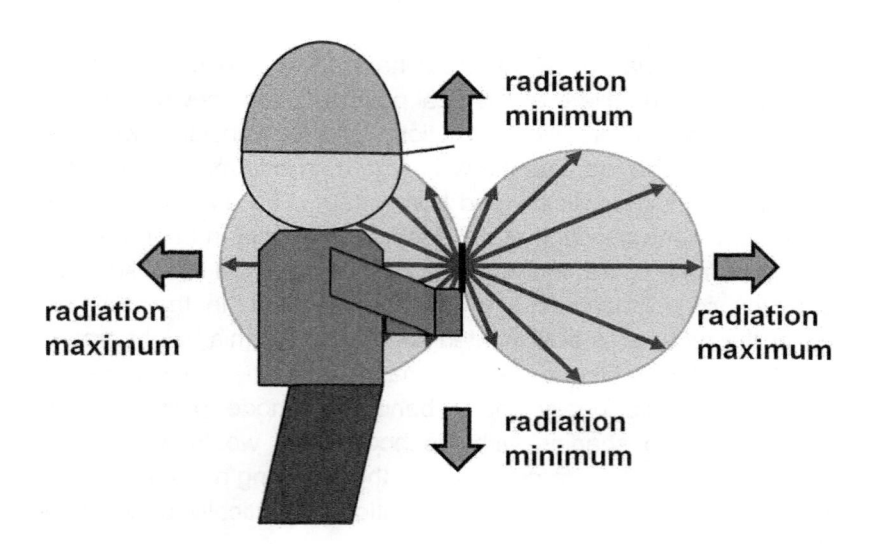

Figure 23: Antenna radiation pattern of a rod antenna

Because of the radio shadow by the pilot, the antenna beam is in practice smaller to the rear than to the front. The case shown in the figure with equally sized antenna beams forwards and to the rear would only occur if the antenna were completely freestanding. As the radio-controlled model is usually in front of the pilot, the radiation to the rear is not as important. The flat illustration in the diagram may also not fulfill the real three-dimensional conditions. You have to imagine that the gray area continues around the antenna, so in reality it has the shape of a donut or a floating ring. Just as the antenna radiates to the front or back, it also radiates to the side, i.e. to the right or to the left.

Antenna gain

The term 'antenna gain' is especially important in the comparison between the different antennas used. There are also differences between the MHz and 2.4 GHz RC. However, to explain it accurately, a little more must be explained at this point.

One type of theoretically used antenna is the isotropic radiator according to Figure 24. Theoretically means that such an antenna is not built in practice, but is only used for comparative purposes with real antennas. This is a punctual antenna, which is believed to have a specific transmission power. The radiation pattern is a ball, because a punctiform antenna radiates in all directions evenly. When the rod antenna with the vertical orientation from Figure 28 was discussed above, the radiation had a maximum in the horizontal direction. Thus, this antenna has a main direction in which it emits better than in other directions. In contrast, the isotropic radiator has no major direction in which it emits better than in others. The recipient does not have to be in the horizontal direction, but can be anywhere, and the same and equally strong signal would be received at the same distance.

To make a comparison between a rod antenna and an isotropic radiator with respect to the strength of the received signal, it is first assumed that both transmitter antennas with the radiation characteristics of Figures 23 and 24 radiate with the same total power. Then it is logical that a receiver which is placed in the horizontal direction around the rod antenna must receive a stronger signal than if it was sent out by an isotropic radiator. This advantage must be paid by the fact that the receiver would no longer receive any signal if it were placed in the direction of the rod antenna, because there the radiation of the transmitter antenna is at a minimum. The ratio between the strength of the received signal which originates from the rod antenna in the direction of maximum radiation and the strength of the received signal coming from an isotropic antenna is called antenna gain. It is further assumed that the receiver is located at the same distance to the transmitters.

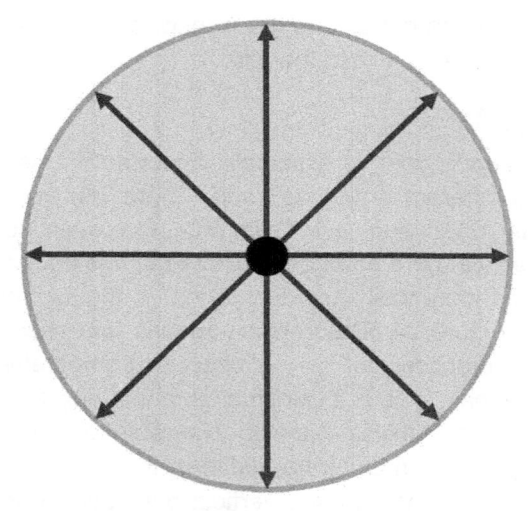

Figure 24: Antenna radiation pattern of an isotropic radiator

So it's always a compromise: more antenna gain in one direction also means that the directional characteristic is very strong and that the antenna emits less in other directions or not at all. One can imagine this with a shower head. If you hold some of the water holes closed, so the shower splashes the water with the remaining holes in a more directed manner and further.

It was mentioned above that the length of the antenna always has the same dimension as the wavelength. The calculation of Example 2 in Section 3.1 shows that at 72 MHz the electro-magnetic waves have a length of 4.17 m. For 27 MHz it would even be longer. The antennas used for this MHz RC are, however, for practical reasons, shorter. The receiving antennas are much shorter. Only a few centimeters long, the 2.4 GHz antennas can however in practice always be tuned optimally to the wavelength. The antenna gain of a practical 2.4 GHz antenna is therefore higher than that of a 72 MHz antenna. This also means that a 2.4 GHz system compared to a 72 MHz system generally needs lower transmitter power to achieve the same signal strength at the receiver.

4.3   Rod antenna and its orientation

Up to now all considerations of the basics and the practical realization of antennas have been made on the basis of the rod antenna. This type of antenna is very often used in radio control systems for model construction. The antennas are usually designed to be rotatable and foldable, so they can be rotated and swiveled in almost any direction. This can also be seen in Figure 25, in which a typical rod antenna is shown.

Figure 25: Radio control system with rotatable and foldable rod antenna

The antenna radiation pattern of the rod antenna was presented and discussed in Figure 23. It is important for the orientation of such antennas that the maximum radiation is perpendicular to the antenna. Thus, when the antenna is vertically oriented, the radiation in the horizontal direction around the antenna is at a maximum. Many model pilots and captains do not worry in practice about the position of the antenna. They simply choose an arbitrary orientation, maybe angled slightly from the transmitter. This is generally not a bad idea, because the distance and direction to the models will change constantly.

But there are preferred directions for different types of models. They are designed to ensure that the antenna is oriented optimally to the model in a variety of cases.

Antenna orientation for model airplanes and helicopters

In flight, model airplanes and helicopters are seen mostly obliquely upwards from the pilot. Therefore, the antenna is slightly folded back toward the head compared to Figure 23. Thus, a maximum radiation in the direction of the model is achieved, just perpendicular to the antenna. Today there are different views on whether electromagnetic waves are harmful or not. If the tip of the antenna points directly at the head of the pilot and thus has a minimal radiation there, that is optimal in this respect as well. Even for a landing, where actually a vertical antenna position according to Figure 22 would be optimal, the antenna still radiates well enough in this orientation.

However, in this case it is important to note that the optimal receiver antenna orientation is also vertically, because of the polarization. This will be discussed later.

Antenna orientation for car and ship models

For car and ship models, the antenna can be oriented either completely vertically or slightly inclined forwards. Here too, due to the vertical polarization, a vertical orientation of the receiver antenna is necessary.

Receiver antenna

At the receiver, the antennas are almost always wire pieces with a certain length. As with the transmitter antenna, this is also at the receiver of the same dimension as the wavelength. Without the feed line it is with 2.4 GHz receivers for example $\lambda/4$ or about 3.1 cm. For the 72 MHz receivers it was usually much longer. Therefore it was always a special challenge to install the cable in the models.

These wires are also rod antennas. So they also have an antenna radiation pattern which is analogous to Figure 23. Thus, the receiving antenna needs to be aligned vertically, the same as the transmitter antenna, so that a maximum receiving signal is obtained. So for vertical transmitter antennas the receiver antenna should also stand out vertically from the model.

To illustrate this somewhat, Figure 22 is represented in a different way. It shows perfectly oriented transmitter and receiver antennas in the vertical direction. If one imagines that the transmitter is a light spot, then this light spot has an optimum projection on the entire length of the receiving antenna. Figure 26 shows this.

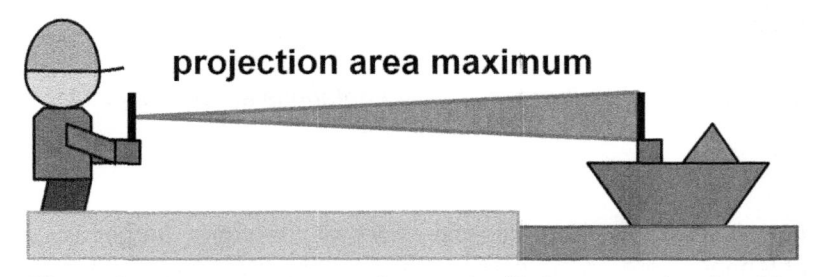

Figure 26: Optimal antenna alignment with the example of a ship model

Although this representation is somewhat simplified, it shows the true conditions pretty well. If the receiver antenna is now turned a little compared to the transmitter antenna, as shown in Figure 27, then the projection becomes smaller, and so the received signal is somewhat weaker.

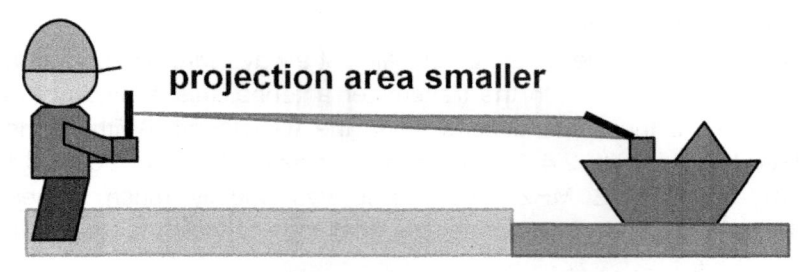

Figure 27: Slightly turned receiver antenna

If the receiver antenna were not rotated in this direction, but figuratively speaking into or out of the page, the received signal would also be weakened due to the different polarization.

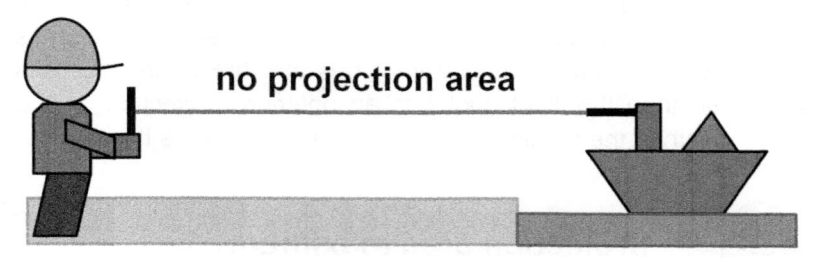

Figure 28: Transmitter and receiver antennas are perpendicular to each other

Figure 28 shows the extreme case with mutually perpendicular antennas. Here no projection exists and therefore there is actually no received signal. Here, however, are already the limitations of this simplified representation. In practice, there is still a reception signal present at this antenna orientation. The transmitter antenna radiates not just into the drawn direction, but according to the antenna diagram of Figure 23 also slightly weaker into the other directions. Thus, there are also electromagnetic waves, which hit the ground and arrive reflected at the receiver. In practice, therefore, there is still reception. Nevertheless one should try to avoid that the antennas are perpendicular to each other.

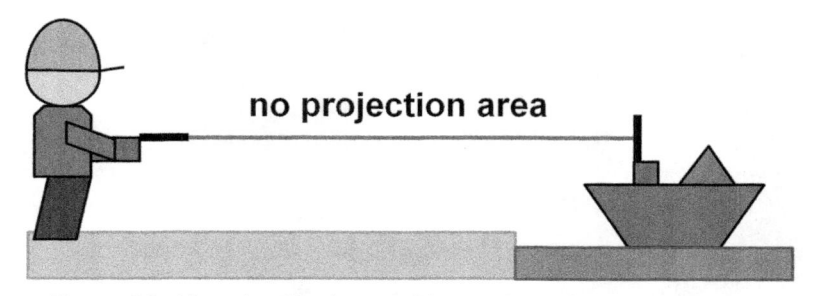

Figure 29: Here too the transmitter and receiver antennas are perpendicular to each other

Figure 29 shows the same case as Figure 28. The receiving antenna has the same antenna radiation pattern as the transmitter antenna. If it is perfectly aligned to the transmitter as here, one can imagine a point of light here also. If this has no projection on the poorly oriented transmitter antenna, the signal is also poor. In summary, these findings can be summarized as follows:

A transmitter rod antenna should be perpendicular to the direction of the model; in this direction the radiation is at its maximum. It should never directly point to the model, otherwise the signal is minimal. The same applies for the receiver antenna. It should be perpendicular to the direction of the transmitter. It should never point in the direction of the transmitter.

An example for car or ship models
These models are usually radiated from above, as has been shown for example in Figure 16. Therefore, the receiving antenna is here often mounted so that it protrudes upwardly from the model. Figure 30 shows the example of a car model. In this case, the optimum orientation of the transmitter antenna is also vertical, i.e. upwards. Normally, in this way there is also visual contact between the transmitter and receiver antenna.

Figure 30: Antenna of a car model

4.4 Patch antenna and its orientation

A completely different type of antenna is the patch antenna. This has a greater directional characteristic than the rod antenna. It typically consists of a plate with the length of $\lambda/2$. Behind it is a dielectric plate, i.e. a non-conductive layer, and then again followed by a conductive layer. This acts as a reflector and shields the electromagnetic radiation to the rear. That is the reason for the strong forward directivity. The antenna gain is therefore larger than with rod antennas. The power of the high-frequency generator must therefore be reduced somewhat to achieve the same signal strength at the receiver as with a rod antenna. Figure 31 shows a transmitter module with patch antennas, which was mounted on a universal radio control.

For the orientation of the antenna, the antenna radiation pattern of Figure 32 should be used. For an optimal receiving signal, the front of the plate must therefore point in the direction of the model. For model airplanes and helicopters an orientation slightly inclined upwards would be meaningful as shown. For car or ship models, the plate should be directed approximately straight forwards. With the patch antenna, it is advantageous if the model pilot always turns into the direction of the model together with his radio control transmitter.

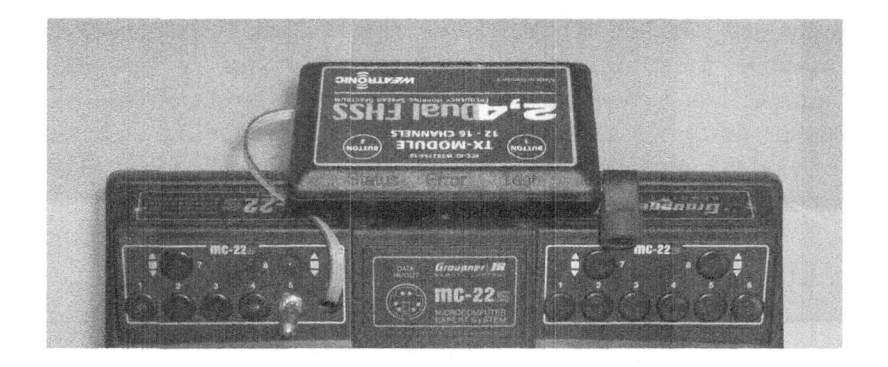

Figure 31: Transmitter with patch antennas

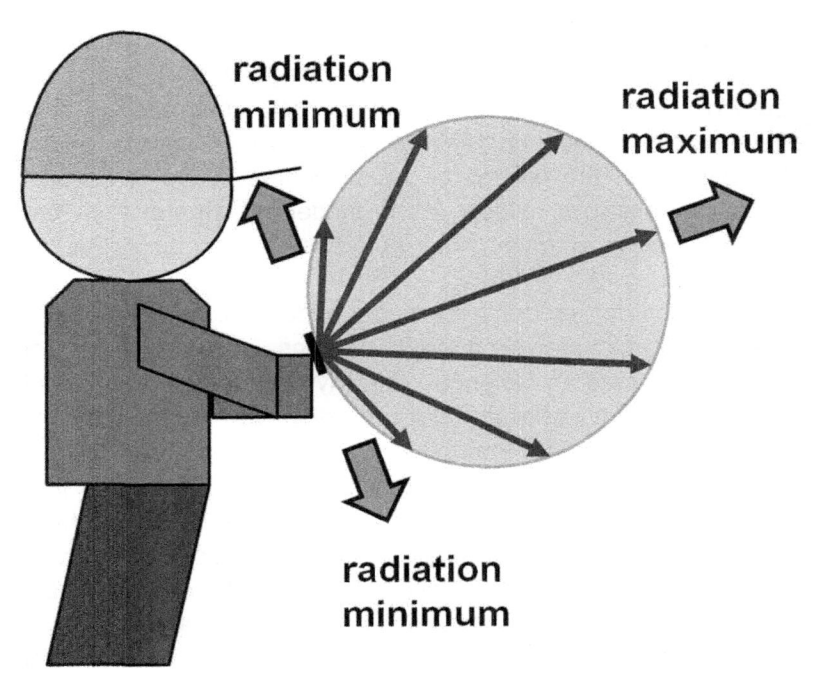

Figure 32: Antenna radiation pattern of a patch antenna

Which of the presented antennas is best suited for use in the 2.4 GHz radio controls can't be explained conclusively, since they each have their own special properties as discussed above. For this reason, they all have a reason for existing. The rod antenna with less directionality and lower antenna gain is slightly less sensitive to the orientation, but its high-frequency generator requires more power to achieve the same strength of received signal. As discussed above however, safe reception depends not only on the strength of the received signal. It depends above all on whether a signal is still received by the model even if it is not standing optimally to the transmitter antenna, or when reflections on the ground occur or when obstacles shade the signal.

4.5 Diversity

The examples of vertical transmitter and receiver antennas as shown above are useful for model cars or model ships. The radio-controlled models, however, are all designed to move during operation. That is also the reason why radio control exists at all. Model cars and model boats move in only two dimensions, so they always stay on the ground or on the water. There is no case, except perhaps in a rollover of the model car, where a vertically adopted receiver antenna takes up a different orientation than this one.

The situation for model aircraft or model helicopters is totally different; they can move in three dimensions and in any orientation. Therefore, a receiver antenna optimally positioned for a straight flight can suddenly take any desired orientation together with the model. Imagine the extreme case of a 3D model helicopter, which is controlled by an experienced pilot. It moves so quickly in all directions and then changes orientation so that it can barely be tracked by a spectator. Even with aerobatic models, the orientation of the antenna is constantly changing. It goes without saying that in this case the optimal reception can't be guaranteed at all time.

But especially with airplanes the safe reception at all times is very important because even a very short dropout could lead to a total loss of the model.

Two antennas as the characteristic

For this reason, multiple antennas with different orientations are mounted, especially with model planes and model helicopters, but often also with other kinds of models. For this the term 'diversity' is used.

The basic idea of the diversity function is that these antennas and their electronic circuits can then process transmitter or receiver signals of different polarizations. The best possible signal is then always switched to. Since the model together with its receiver usually turns more than the transmitter, the first diversity systems were developed for the receiver units. However, since it depends only on the relative orientation between the transmitter and receiver antennas, the effect is the same in principle whether the receiver or the transmitter, or both, have a diversity function. For this reason, the diversity function can be applied to both.

Receiver

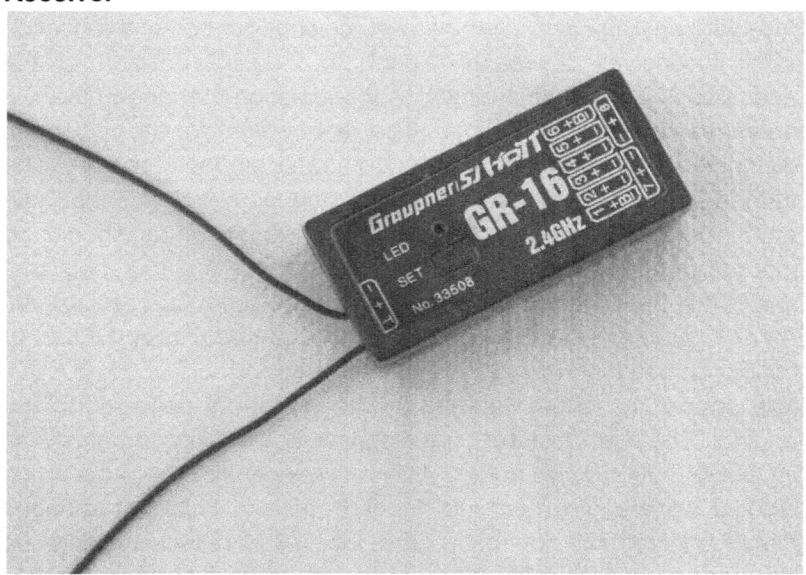

Figure 33: Receiver with diversity function

Figure 33 shows a receiver with diversity function. Diversity could in the simplest case just mean that the receiver switches between

55

two different antennas and evaluates the better signal. However, there is the possibility that in addition to the antennas also the whole internal part of the signal processing is double. This variant is a bit more complicated to implement, but also safer. The more subsystems that are double – also called redundant in technical language – the greater the reliability. Of course, there are also intermediate variants between these two extremes. So it is possible that only the high-frequency part including antenna is doubled, but the signal processing is only available once.
To answer the question of the optimal orientation of the two antennas, the following mnemonic will be given and discussed.

In diversity mode, the two antennas are oriented at 90° to each other in the optimal case. The antenna ends should be as far away as possible from each other. Antennas must not be shortened or extended, but they should be used as stated by the manufacturer.

The 90° requirement is formulated based on the polarization. If one receiving antenna is rotated away somewhat out of the polarization axis, it receives a weaker signal. In the 90° position of the antennas is it ensured in almost any orientation that one of the two antennas is receiving a signal. The requirement for the maximum distance of the antenna tips is due to a possible radio shadow. The further apart they are from each other, the less likely it is that both antennas are affected simultaneously. The antennas should be mounted outside of the model, because there are materials that shield the 2.4 GHz waves and contribute to the loss of reception. As an example especially carbon-fiber-reinforced plastic (CFRP) or carbon is mentioned here.
The supply line should always be left as it was delivered in the original. The manufacturers have optimized the receiving antennas including their supply cable. 2.4 GHz systems are due to the much smaller wavelength much more susceptible to improper changes than the MHz radio controls. With a long 72 MHz receiver antenna the reception was still there even when this was somewhat shortened or lengthened. At 2.4 GHz even a change in the length of a few millimeters can cause a complete signal loss.

Transmitter

Figure 34: Transmitter with diversity function

Figure 34 shows a transmitter with diversity function. Exactly the same as for the receivers, there are different versions available. In the simplest case, only two antennas are coupled to the same high-frequency part. With even more reliable systems, such as the one shown in the image, two completely independent transmission units are installed. For the same reason as above, it is also reasonable that the two antennas are oriented angled to each other. In addition, when also the receiver is equipped with diversity, two independent transmission paths can even be realized.

Use of diversity systems in the 2.4 GHz RC

In summary it can be said that the diversity function is now used much more frequently with the 2.4 GHz RC than it was earlier the case with the MHz RC. With airplanes and helicopters, the author even recommends the use. This is certainly due in part to the technological leap. Today, computer chips for the analysis or the control of diversity systems can be integrated much more easily and cheaply into the transmitter and receiver than was previously the case.

On the other hand however, the 2.4 GHz systems are slightly more susceptible to sudden reception problems compared to the MHz

RC. One reason is the signal loss in the moving models or the radio shadow discussed in Section 3.4. With 2.4 GHz technology it is slightly larger due to the wavelength, which is more similar to light. Another reason can be found in the receiver antennas themselves. The MHz antennas were much longer and were in the example of a model plane usually first placed inside the hull, then clamped to the fin, and the last part was then hanging down. So it was almost always ensured even without a diversity system that at least a part of the receiver antenna length was somewhat optimally oriented. Thus, the waves emitted by the transmitter could be captured.

From the perspective of the author, however, the 2.4 GHz technology compensates for it because of the gained confidence that you no longer have to come to an agreement over the channel, as is the case with the MHz technology. Nevertheless, the following mnemonic will be at the end of the chapter:

For safety reasons, model planes and model helicopters at 2.4 GHz radio controls should always be operated with diversity systems. For model ships or model cars, the use of diversity systems is at least recommended.

Findings in brief

As in the last chapter, here the main findings will be presented again in brief.

For optimum reception, the transmitter and receiver antennas should have the same orientation (polarization, antenna radiation pattern of Figure 23).

A transmitter rod antenna should be perpendicular to the direction of the model, as in this direction the radiation is maximum (Figure 26). It should never directly point to the model, as otherwise the reception signal is minimal. The same applies for the receiver antenna. It should be perpendicular to the direction of the transmitter. It should never point in the direction of the transmitter.

The stronger the directivity of the antenna, the greater the antenna gain and the smaller the power required for reliable reception in this direction (Figure 23, antenna radiation pattern, antenna gain).

For safety reasons, model planes and model helicopters with 2.4 GHz radio controls should always be operated with diversity systems. For model ships or model cars the use of diversity systems is at least recommended.

In diversity mode, the two antennas are oriented at 90° to each other in the optimal case. The antenna ends should be as far away as possible from each other. Antennas must not be shortened or extended, but they should be used as stated by the manufacturer.

5. Modulation and data transmission methods

Figure 35 shows the various functional blocks of a 2.4 GHz radio control transmitter. Figure 36 shows the same for the receiver. At the beginning of Figure 35 is the stick, and at the end of Figure 36 is the horn of the servo, which finally converts the motion given by the stick. The last two chapters (3 and 4) actually describe only the wireless transmission between the transmitter and receiver antenna, which is what happens between Figures 35 and 36. This chapter describes the function of the blocks shown in the figure.

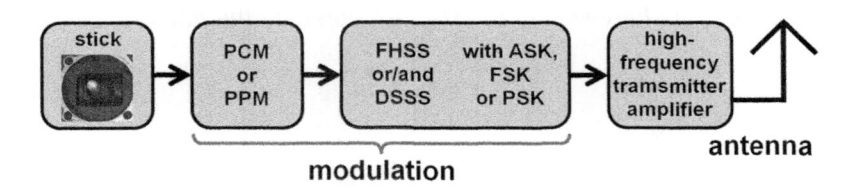

Figure 35: Block diagram of a radio control transmitter

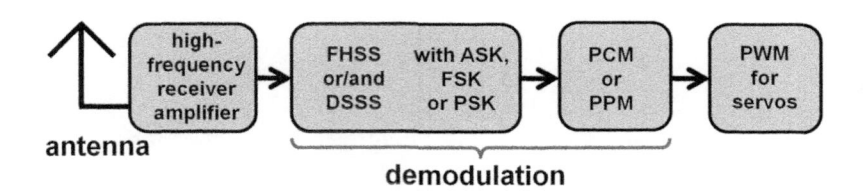

Figure 36: Block diagram of a radio control receiver

So for the transmitter the question of what happens between the stick and the antenna will be discussed. For the receiver, we consider what happens between the antenna and the servo. Between the stick and the high-frequency amplifier of the transmitter, there are two blocks which are named 'PCM or PPM' or 'FHSS and/or DSSS using ASK, FSK or PSK'. They all fall under the generic term 'modulation'. Accordingly, these blocks are found in reverse order in the receiver. There, they fall under the generic term 'demodulation'. Separate sections are now dedicated to these blocks. In Section 5.1 'PPM and PCM' is treated and in Section 5.2 'ASK, FSK and PSK'.

Code division multiple access (CDMA)

The terms FHSS ('frequency-hopping spread spectrum') and DSSS ('direct sequence spread spectrum') also belong to modulation. However, since they represent important fundamentals of the 2.4 GHz transmission, they are each in a separate section, namely Sections 5.4 and 5.5.

The generic term for this is 'code division multiple access' (CDMA). The point is that different transmitters can transmit data to their receivers without disturbing the others. This not only means the radio controls, but everything which is transmitting or receiving on the ISM band from 2.40 GHz to 2.48 GHz, for example WLAN or Bluetooth.

5.1 PPM and PCM

"What is modulation actually?", you might ask yourself at this point. Wireless communication works well only at high frequencies in the MHz and GHz range. With a signal transmission with lower frequencies, the transmitter power would be very high for physical reasons. If you want to transmit information wirelessly, you have to link it in a meaningful way with such a high-frequency signal in the MHz or even GHz range. The information is also called desired signal and the high-frequency signal is the carrier signal. This process of meaningful combination of desired signal and carrier signal is generally called modulation. The combined signals are then amplified and transmitted via the electromagnetic waves, as discussed in the last two chapters. They generate a signal at the receiver. There it is amplified first and still contains the desired signal and the carrier signal. Through the reverse process, namely the decoupling of the desired signal from the carrier signal, the desired signal is recovered. It can then be used to control servos, motor controllers or other peripherals. The decoupling of desired signal and carrier signal is called demodulation.

That is basically simple, one might think, but modern technology offers many sophisticated options for modulation or demodulation. Thus, an entire chapter is necessary to explain just those which are used in 2.4 GHz radio controls. The variety of these types of modulation is also the reason that 2.4 GHz technology could prevail in wireless data transmission.

Pulse pause modulation (PPM)

In order that the information about the stick position can be transferred to the receiver, it must first be brought into a specific signal form. While the first radio control systems did not yet have a certain standard in this respect, a little later a variable pulse length between 1 ms and 2 ms after Figure 37 was established.

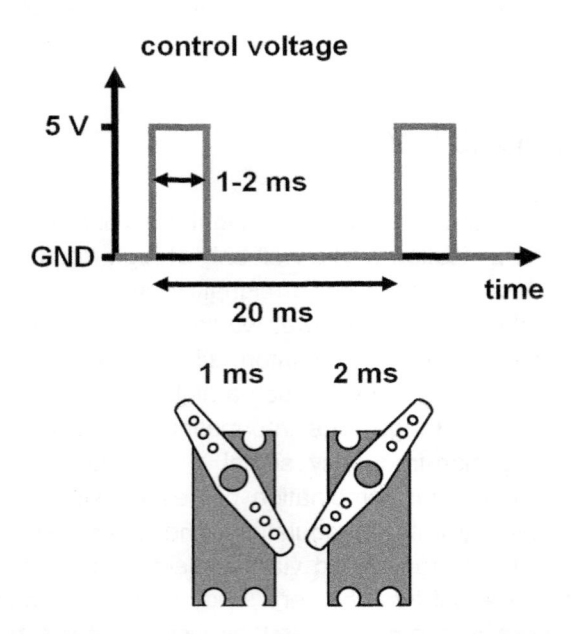

Figure 37: Control signal and associated servo positions

A 1 ms or 2 ms long signal means that the stick is in the one or the other extreme position. Thus, this means that the servos on the model side are also in the one or the other extreme position. A 1.5 ms long signal therefore means that the stick and the servo are in the middle. These signals are also called PWM or pulse width modulated signals.

First, the electronics in the transmitter evaluate the position of the stick and convert it into this PWM signal. Then, the PWM signals of all the channels are connected in series. This is illustrated in Figure 38.

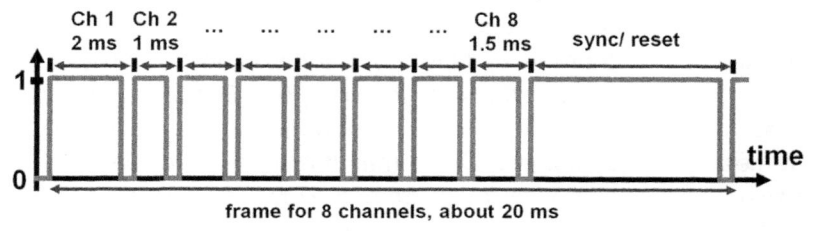

Figure 38: Frame of the concatenated PWM signals of eight channels

The signal thus formed is also called 'frame' or 'summed signal'. In this example, the 1 ms to 2 ms long PWM signals of eight channels are shown. As examples the single servos would get the following control signals: servo of channel 1: right (or left) position, servo of channel 2: left (or right) position, the servos of channels 3–8: middle position. This is followed by a long reset or synchronization signal. The electronics in the receiver must indeed count up the channels and assign them the appropriate pulse length. The long signal at the end of the frame indicates that the signal sequence will then be repeated again starting with channel 1. But in most radio control systems the signal in Figure 39 is used.

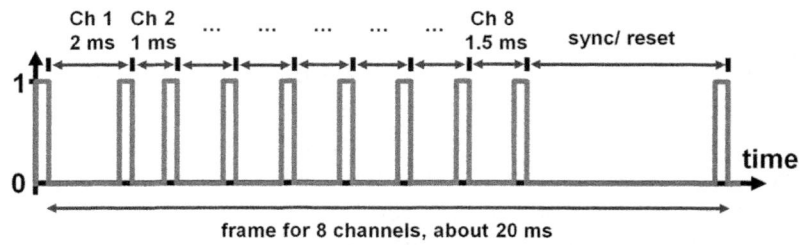

Figure 39: Pulse-pause modulation

It is the inversion of the signal from Figure 38. This means that in each case it is '0' when that of Figure 38 is '1', or '1' when that of Figure 38 is '0'. Now the term 'pulse-pause modulation' is also better understandable. The information of the signal is not included in the pulses of the same length, but in the pauses between these pulses. A frame with eight channels is so repeated approximately every 20 ms. Radio controls that work with PPM update the position of their servos about 1/0.02 = 50 times per second. PPM is for a very long time a signal standard for radio controls. On the question of whether this update rate is sufficient for all applications in model construction, the experts disagree. Some say that the reaction time of humans is much slower than the repetition of frames in 20 ms intervals. Some manufacturers have reacted and developed systems with faster frames. However, this only works with digital servos. It should be noted here that there are two definitions regarding PPM. Namely PPM is also the abbreviation for the pulse-phase or pulse-position modulation which is used in telecommunications. Therefore, the shortcut often leads to confusion. In this book, however, the pulse-pause modulation discussed above is always meant when the abbreviation PPM is used.

Pulse-code modulation (PCM)

With the PPM signal the length of the pauses between pulses is relevant. The receiver evaluates them for all channels and controls the servos. Thus, it is important upon receiving that the length can be accurately determined. However, exactly this is a small disadvantage of this signal standard. Both in the generation of the signal in the transmitter as well at the signal transmission and further in evaluating in the receiver, small errors are introduced. Thus, the servo position doesn't always correspond exactly to the stick position. Although the error is very small and has mostly no meaning for practical model construction, manufacturers have looked for ways of representing the stick position with another signal which allows a more precise transmission. To do this, they apply pulse-code modulation (PCM). For this purpose the stick position must first be present in digital form. If the stick position exists, for example, as a voltage, it can be converted into a number using an analog-to-digital converter.

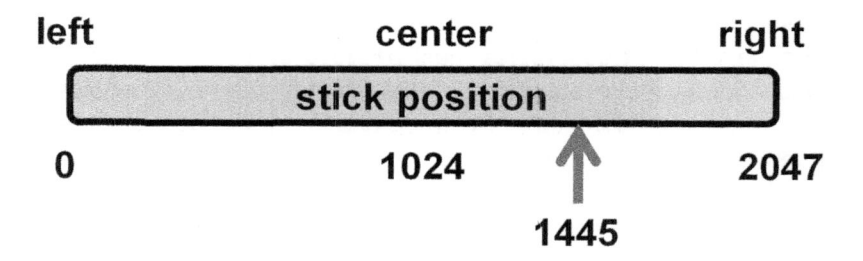

Figure 40: Analog-to-digital conversion of the stick position

Figure 40 shows the example of a so-called 11-bit converter. An 11-bit converter converts the voltage into 2^{11}, so 2x2x2x2x2x2x2x2x2x2x2 = 2048 different values (from 0 to 2047). So if the stick position is for example at the arrow, in the figure about middle right, the 11-bit converter calculates it as the number 1445. Other converters are also used, for example 8- to 10-bit converters with $2^8 = 256$, $2^9 = 512$ or $2^{10} = 1024$ different values. This number must now still be brought into a suitable form so that it can finally be transmitted to the receiver. For this purpose it is coded. Thus, very often binary code is used. Each bit then has a different value. It starts at 1 and is always doubled, in the case of the discussed 11-bit converter up to 1024. This is illustrated in Figure 41.

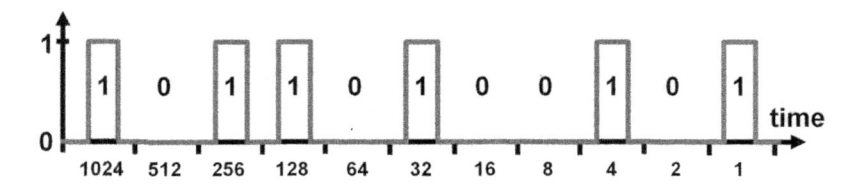

Figure 41: Pulse-code modulation of the number 1445

Now you can add up all the values of the set bits in which that is a '1'. This results in 1024 + 256 + 128 + 32 + 4 + 1 = 1445. This code is unique. This means that any combination of set and not set bits can be assigned exactly to a number between 0 and 2047. This sequence of '0' and '1', so this 10110100101, now replaces the PWM signal of Figure 37.

Assuming that a frame needs to be transmitted with a plurality of channels, the bit sequences must be hung behind the other as it was already the case with the PPM signal. Figure 42 shows a schematic representation of a three-channel radio control as it is used in some car models.

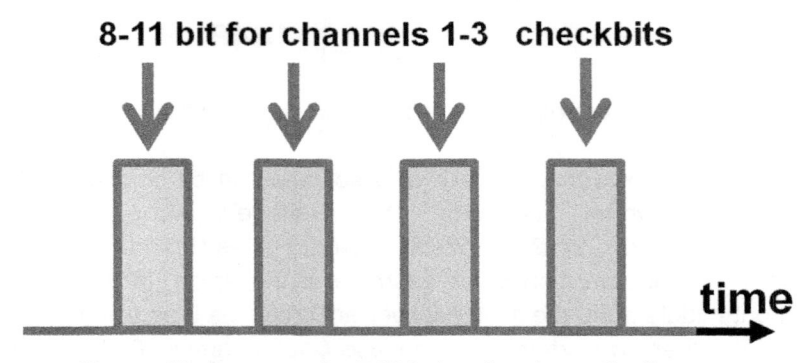

Figure 42: Frames of a PCM signal, schematic diagram

After the bit sequences of the three channels, a check bit is generated here. The bits of the three channels are therefore added. If, for example, 1 and 1 is added, it results in binary in 10. In the so-called binary representation, indeed only zeroes and ones occur. This check bit is then the rear number, so the '0'. When this signal is transmitted, the receiver can also perform this calculation with the received bits.

If its own calculation differs from that of the received check bits, an error must exist in the data transmission. The receiver will then go into the 'hold' or 'fail safe' mode, according to its programming (see Section 2.2).

With demodulation, the receiver must first restore for each channel the number which was coded by the transmitter. Afterwards it again generates a PWM signal for the servos. Check bits in PCM signals is state-of-the-art technology. In the manner of their calculation and how they are integrated into the transmission signal, most manufacturers have their own standards. Figure 42 with the check bits after three channels shows only one of many possibilities for implementation.

Also the length of frames is not the same with all manufacturers. Unlike the PPM signal with a frame which will take approximately 20 ms for the eight channels, the length of the PCM frames sometimes varies considerably. It depends on the one hand on how long the individual bits of the signal are. On the other hand, how many channels are transmitted at all, and how extensive the check bits are, also has a significant impact.

PPM and PCM in comparison

First of all, an often assumed fallacy should be removed from the world. It is namely often incorrectly assumed that PCM only works with digital servos and for analog servos, but that one must always transmit a PPM signal. This is wrong. PCM and PPM is related to the way in which the information is transmitted. Whether digital servos can be actuated is the responsibility of the receiver. This should provide and support the corresponding control signals. As already stated above, it is quite possible that the transmitter communicates to the receiver with a PCM signal, while the receiver provides a PWM signal for a normal analog servo.

To discuss the differences between the two signals, Figure 43 is used. It shows at the top a profile of the control stick on the transmitter. This is first evenly moved from left to right and back again to left. It is further assumed that this movement is performed slowly enough so that the servos get enough time to execute it.

After a transmission using a PPM signal, the movement of the servo is very slightly noisy. So the rudder horn is not exactly in the position in which it should be after the stick position on the transmitter. But for the sake of illustration this is shown greatly exaggerated in the diagram. In practice, the noise and thus the error is considerably smaller than drawn in Figure 43. This is, as discussed above, because the length of individual pulses is not exactly generated, transmitted and evaluated. In reality, the position is always somewhat similar to the stick position.

When transmitting a PCM signal, the position of the servos corresponds more accurately to the stick position on the transmitter. The signal is indeed coded. Therefore, the position of the transmitter stick is reproduced exactly in the evaluation. In contrast to the signal transmission with the PPM signal, with PCM the receiver detects a transmission error, since it evaluates the check bits. In the figure, the behavior of transmission errors has been illustrated in three places. These are in practice also much less than is shown here. When an error is detected, the receiver can pass into 'hold' function depending on the configuration. The servo obtains in this case the same signal until the receiver again evaluates a signal which coincides with its own calculated check bits. In the figure, these are then the edges and in practice this means that the servo stops for a short moment and then jumps to the position corresponding to the transmitted signal.

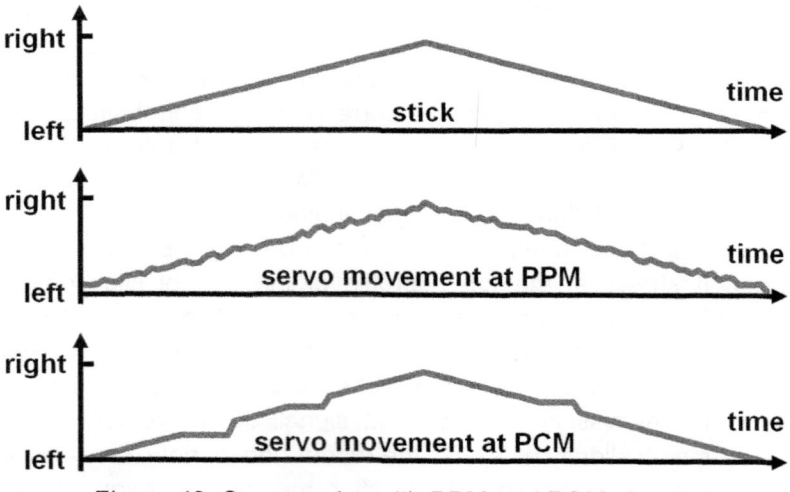

Figure 43: Servo motor with PPM and PCM signals

Regarding the question which of the two transmission signals is better suited for model construction in practice, the experts argue between PPM and PCM. The PPM signal has the disadvantage that the servo position is somewhat noisy, but this servo will always remain approximately in the desired position as long as a usable signal arrives at the receiver. With a PCM signal, the servo position is less noisy. However, in the event of a hold function due to an error, this position differs significantly from the transmitted signal.

PCM is implemented in different ways by the manufacturers. But the PPM signal standard of Figure 39 is always similarly transmitted. Therefore the MHz RC transmitters and receivers of different manufacturers were almost always mutually compatible with the PPM signals.

In order to transmit the signals, however, further modulations in the high-frequency range are always needed. As the next sections show, the transmission possibilities of the 2.4 GHz radio controls are very versatile. So you can almost speak of different transmission philosophies followed by the individual manufacturers. For this reason, the transmitters and receivers of the different models are usually not or generally much less compatible with each other than was the case with MHz radio controls. So as compatibility is not given anyway, it doesn't matter if one uses PCM right from the beginning. Thus, PCM signals are used today in most radio controls in the 2.4 GHz technology.

5.2 High frequency, ASK, FSK and PSK

PPM and PCM signals are not however sufficient on their own to transmit a radio control signal. As discussed in Chapter 3.1, electromagnetic waves are generated from sinusoidal alternating voltages at high frequencies. PPM and PCM are so-called baseband modulations. This means that these signals do not have anything to do with the much higher signal transmission with 2.4 GHz, but they only represent a preparation for this. Therefore, the PPM or PCM signals are also called baseband signals. As can easily be seen in Figures 39 and 41, they consist of only two states, 0 and 1. Another name for this is binary. These baseband signals are now modulated with a so-called binary shift keying onto the high-frequency carrier signal. For this purpose, either the amplitude or the frequency or the phase is changed for the transmission. How it works exactly and what methods are used is shown in the three subsections for ASK, FSK and PSK. These three methods are all digital modulation types.

ASK (Amplitude Shift Keying)

ASK is an abbreviation of the term 'Amplitude Shift Keying'. The PPM or PCM signals discussed in the last chapter are shown in Figure 44 as coded signals. With the ASK at logic 1, a sinusoidal AC voltage with a transmission frequency in the ISM band, so somewhere between 2.4 GHz and 2.48 GHz, is now available. With logic 0, there is no signal. The case illustrated here is also called 'on–off keying' (OOK), since the high-frequency signal is either present or not at all. The very first example where OOK was applied was the transfer of Morse code. There is also the possibility that a signal would also be present at a logical 0. This would then have the same frequency as with a logical 1, but with lower amplitude. That would also be ASK.

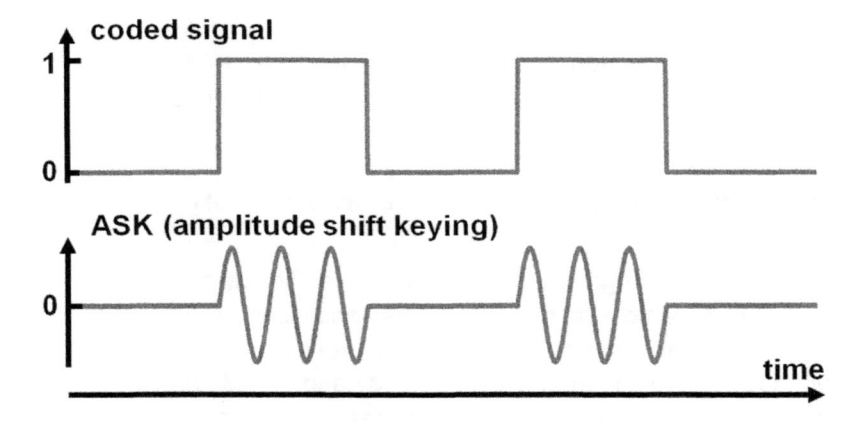

Figure 44: ASK (amplitude shift keying)

ASK is today only rarely used in model radio control. It is treated here only for completeness and as a simple starting point for the explanation of FSK and PSK. One can easily imagine that a radio shadow or reflections may attenuate the signal so that the receiver would recognize a logical 0 rather than a logical 1.

But ASK has a practical relevance in other fields of technology, for example in the transmission of information in optical fiber cables.

FSK (Frequency Shift Keying)

FSK is an abbreviation for 'Frequency Shift Keying'. The two logic states 0 and 1 are transmitted at two different frequencies. These are both also within the 2.40 GHz to 2.48 GHz band. Figure 45 shows the signal course. Again, the coded signal is shown at the top; it could be either PCM or PPM. Below it is shown what FSK makes with it.

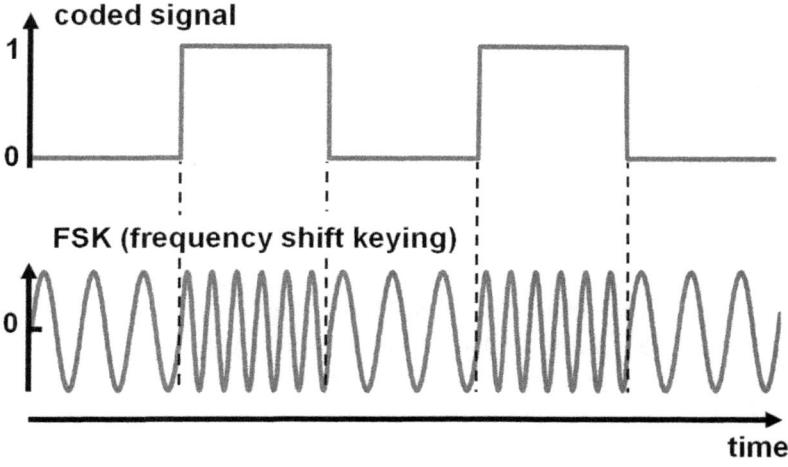

Figure 45: FSK (frequency shift keying)

This type of signal transmission is much less susceptible to interference than the ASK discussed above. By the 'binding', which will be discussed in Sections 5.4 and 5.5, the receiver also knows at any time on which frequencies the transmitter is working at the moment. It must then only distinguish between them. It can detect the 0 and the 1 quite well and is reasonably immune to interferences.

PSK (Phase Shift Keying)

PSK is the abbreviation for 'phase shift keying'. Sometimes also the term BPSK is used, whereby B stands for 'binary'. Figure 46 shows what PSK does with the coded signal.

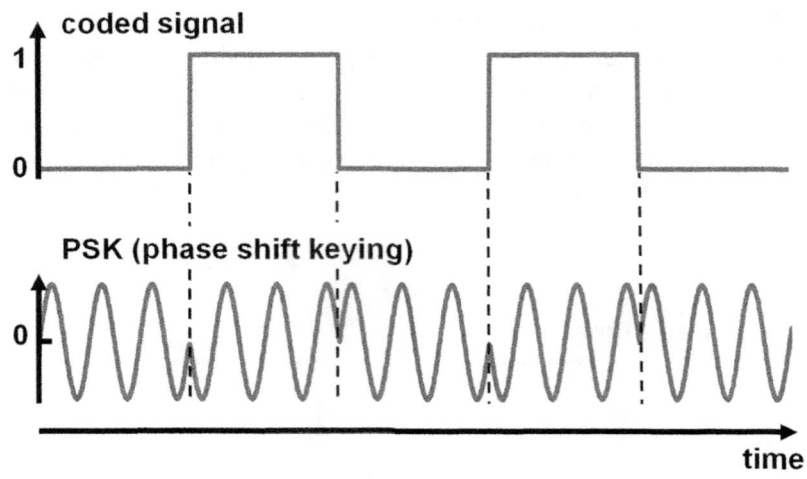

Figure 46: PSK (phase shift keying)

The information here is therefore in the phase jumps at the transitions from logic 0 to logic 1 and logic 1 to logic 0. In the receiver the carrier frequency runs which was agreed by the 'binding' between the transmitter and receiver. It is between 2.4 GHz and 2.48 GHz. In the transmission of a logical 0, the received signal is the same as the carrier signal, and for a logical 1 it is equal to the negative carrier signal. In technical language one also says it has a 180° phase shift. The receiver is fairly resistant to interference and can detect a logical 0 or 1 again and pass it on for further processing in accordance with Section 5.1. Also to be mentioned here is DPSK, which stands for differential PSK. Here the jumps are not transferred in relation to 0 or 1, but due to the change to the previous bit. This results in a phase jump at signal changes from 0 to 1 or from 1 to 0. Thus, the comparison with the carrier signal can be eliminated if the jumps are always detected correctly.

Other shift keying modulations
In practice, with radio control mostly FSK or PSK or methods derived from these basic types are applied. There is, for example, also 4-PSK or QPSK. Q stands for quadrature, where phase shifts

of 90° and 270° will be detected in addition to the phase shifts of 180°. With FSK, there is also 4-FSK. In this case, four different frequencies are transmitted instead of two. Thus not only a coded signal with a logical 0 or 1 can be transmitted, but also a sequence of 00, 01, 10 and 11. This can also go even further by using an 8-FSK.

5.3 Spread spectrum technology

Spread spectrum technology represents the actual core of the transmission of 2.4 GHz. The next two sections focus on two variants – frequency-hopping spread spectrum and direct sequence spread spectrum. Both are used in modern radio-control systems. Although in this book no complicated formulas are generally used, at least one should be described phenome-nologically in order to understand the idea behind spread spectrum. The information quantity of a message can be described as:

Information quantity = bandwidth x signal duration x dynamics

Each component of the formula will be explained at this point with some generally understandable parallels from practice. Since bandwidth, signal duration and dynamics are each multiplied, they are equivalent and can be treated individually.

Bandwidth
Bandwidth is at first a very abstract term. The channels of MHz radio controls are counted in 10 kHz steps. In 2.4 GHz technology the ISM (Industrial, Scientific, Medical) band is available. This includes the frequencies from 2.400 GHz to 2.483 GHz. So the bandwidth of a channel is 10 kHz (10,000 Hz) for the MHz radio controls and for the 2.4 GHz radio controls it is 2.483 GHz – 2.40 GHz = 0.083 GHz = 83 MHz = 83,000,000 Hz. So much more bandwidth is available in the entire ISM band than has been the

case for the MHz radio control. Thus, according to the above formula, a much larger quantity of information can be transmitted.

As an example the transmission of Morse code is mentioned here, in comparison with the transmission of human language. A Morse signal is compared in Section 5.1 to an ASK or OOK (amplitude shift keying or on–off keying). Since it actually consists of an interrupted sinusoidal signal on a single frequency, it is quite narrowband. It therefore leads according to the above formula to a relatively low information quantity. The human language works in a wider frequency range. The phone transmits signals from about 300 Hz to 3,000 Hz. In this range the human language is very easily understandable. The bandwidth is thus 3,000 Hz − 300 Hz = 2,700 Hz, so this bandwidth is significantly higher than in Morse code. Thus it leads to a larger quantity of information according to the above formula.

Signal duration

The above example can now be completed with the signal duration. Like the bandwidth, this is also a factor in calculating the information quantity in the above formula. In order to transfer the same information quantity, the signal must last longer if it is transmitted with narrowband Morse code than if it is transmitted by the wider band human language. That makes sense, because a sequence of words with the information quantity 'Hello World' can be said within about one second. If it is transmitted in Morse code, it takes several seconds, so it's much longer.

Summarized, this results in the following mnemonic:

With Morse code, the bandwidth is narrow, so the signal duration must be longer to transfer a certain information quantity.
With the human language the bandwidth is large, so the signal duration will be shorter for the transmission of a certain information quantity.

In the context of spread spectrum technology this means that
We should take advantage of the whole range from 2.40 GHz to 2.48 GHz!

Dynamics

The last factor in the above formula, namely dynamics, is missing. It is a measure of how big the power difference is between the transmitted signal and the noise which is always present. Since the noise is assumed for simplicity to always be of a similar size, the dynamics is thus a measure of how large the power of the transmitted signal is.

This can also be shown well by an example from practice. In the above formula, the signal duration will remain the same. Less dynamics, which is less power, will be compensated with more bandwidth. This is shown in Figure 47.

Figure 47: Top: narrowband, much power per truck; bottom: broadband, low power per car

Once again, here the formula 'information quantity = bandwidth x signal duration x dynamics' is explained, this time based on traffic flow. The information quantity here are the four boxes, which are at the top transported together on a truck and below distributed across four small cars. In the upper and lower image the same information quantity is being transported in total. The signal duration is the time during which vehicles are driving from the left to the right edge. With a restriction on speed of 80 km/h, the vehicles need the same time in both the upper and lower images. The bandwidth is represented by the width of the roadway. At the top it has only a single-lane road, which therefore corresponds to a narrow bandwidth. On the bottom it has a four-lane highway, which therefore represents the large bandwidth.

The dynamics, or simplified the power, is represented by the vehicles. Above, the (narrowband) one-lane road is traveled by a truck, which simultaneously transports all four boxes with a higher power. Below, the (broadband) four-lane highway is traveled by four cars. Each of them transports with low power just one box. This is spread spectrum technology!

Figure 48 shows this relation in a somewhat more technical illustration. It is a power/frequency graph, which is often seen in combination with spread spectrum technology.

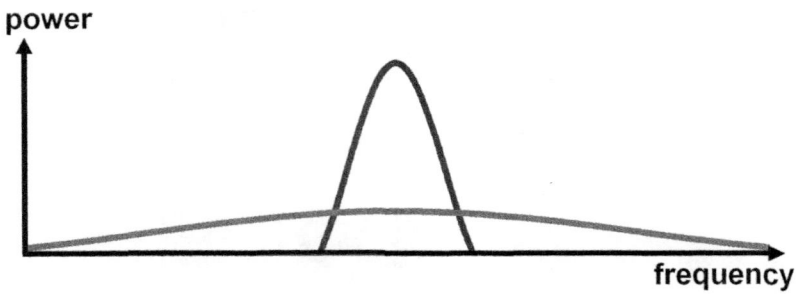

Figure 48: Power/frequency graph

The truck which transports the four boxes together with a higher power on a one-lane road corresponds to the narrow and higher curve. The four cars which carry their boxes, each with low power on the four-lane highway, correspond to the wider and flatter curve. This wide and flat curve is spread spectrum technology. Strictly speaking, the total power of the two curves is the same overall. With the flat curve, it is simply distributed on a larger frequency range. But with the flat curve the power density is smaller. It does not have any power peak like the narrowband curve, but it is wider and less high.

This can also be described using the example of traffic flow in Figure 47. Above, the road is narrow and the only truck has a high power; on the bottom the road is wide and the cars have a smaller power. However, if you add up the power of the four cars, this is overall the same size as that of the truck.

Described with a short mnemonic this means:

Spread spectrum technology means having less power density, but more bandwidth.

Properties of spread spectrum technology

Now it is clear why this technology is called spread spectrum. As shown in the figure, the frequency range is made broadband, which means spread. In return the power at all frequencies is smaller, which means the power density is smaller.

Above, the relatively wideband human language was compared with the narrow-band transmission with Morse code. Once again that means based on Figure 48 that the human language would be compared with the wide and flat curve and the Morse code transmission with the high and narrow curve. Since the power density of the flat curve is rather small over the entire frequency range, this curve, i.e. the spread spectrum technology, could also be compared with voice communication in a whisper.

If one sticks with the comparison to the whisper, other interesting properties of the spread spectrum technology can also be discussed. For example, if one wishes to transmit a signal with Morse code and at the same time an interference signal is present that emits at approximately the same frequency as the Morse signal, then the transfer of this message is not possible with this transfer method. One could also say, according to Figure 47, that the truck overturns with its load of four boxes and they never reach the end of the road. When transmitting the signal with a whispering voice, the signal only interferes with part of the message, or according to Figure 47 only one of the cars overturns with its box, while the rest of the boxes still arrive at the end of the road. The information can then probably be reconstructed anyway. We compare the example with listening to a whisper with hearing loss, tinnitus for example, where a permanent whistling sound is perceived.

Up to this point the question why one can easily turn on the 2.4 GHz radio controls without any frequency agreement is still unclear. But the answer is very close! On the ISM band from 2.40 GHz to 2.48 GHz, many modern communication systems are transmitting their data. Taking the comparison with the whispering voice, imagine that many pairs of people are in a room, and each of these pairs whisper in a different language – in German, in English or in Chinese. The communication works quite well. The Germans focus on the German voices, the English focus on the English voices and the Chinese focus on the Chinese voices. With respect to radio controls that means nothing else than – binding! The receiver is adjusted to the language of the transmitter.

These were some comparisons of spread spectrum technology with examples from everyday life. The next two chapters will show how it is actually technically implemented in the 2.4 GHz radio controls.

5.4 Frequency-hopping spread spectrum (FHSS)

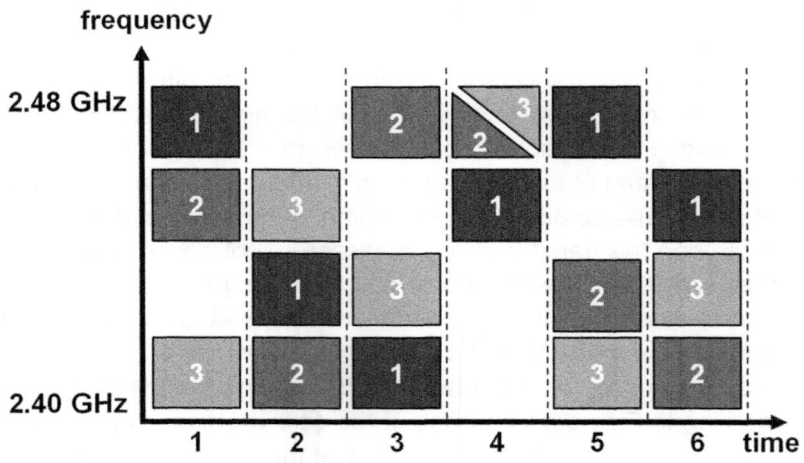

Figure 49: Frequency-hopping spread spectrum (FHSS)

FHSS is often represented in the literature as shown in Figure 49. The four-lane highway according to Figure 47, which figuratively represents a wide bandwidth, is compared here in the technical sense to the frequencies of 2.40 GHz to 2.48 GHz. Once you look at the frequency sequence with number 1, the radio control works at the beginning, that is in the first column, in a frequency range in the vicinity of 2.48 GHz, then it operates in the second column at a lower frequency, such as 2.42 GHz, then at about 2.40 GHz. The frequency range is hopping in a fixed sequence through the entire frequency range of the ISM band.

During the binding process, the transmitter tells the receiver the sequence of frequencies. When the sequence reaches the last column, it jumps back to the beginning, so to the first one. In practice, there are of course many more boxes than shown in the figure. On the one hand, the sequences are actually much longer than six columns. On the other hand, on the frequency axis there are also many more than just the four frequency ranges shown. Figuratively, these represent here the four-lane highway. With practically realized radio control systems, there are usually at least 30 to 300 different frequency ranges. Data is only sent to and received from them for a very short time, in the millisecond range. Then the transmitter and the receiver simultaneously change the frequency.

Technical implementation

With the radio controls which are used nowadays, there are a variety of ways that FHSS can be implemented. One example will be discussed here. Often, the coded signal is first modulated with FSK according to Figure 45. A frequency synthesizer will then ensure that these signals are fed with the FHSS sequence to the high-frequency amplifier. Then they are transmitted to the receiver. This makes in this case first dehopping with the same sequence to an FSK signal and then an FSK demodulation into a coded signal. It will then be converted to the PWM signal for driving the servos.

Parallel operation of multiple radio controls

The boxes with the number 2 represent the sequence of frequencies of another radio control. At the binding the transmitter also communicates this to its receiver. It is visible that the radio controls with the sequences 1 and 2 do not transmit and receive at any time on the same frequency. So the two transmitters and their receivers associated by the binding work at any time without interference.

It is somewhat more specific, if there is an additional sequence 3 with a third transmitter and receiver. Now in the fourth column both the transmitter 2 and the transmitter 3 are sending randomly at approximately 2.48 GHz. Both receivers will now not receive the correct signal, since the two transmitters interfere with each other. In contrast to MHz radio controls, with 2.4 GHz technology with an FHSS this is not tragic at all. In the MHz radio control the two transmitters with the same channel would always send at the same frequency and so a receive problem would be present at all times. As mentioned above, the radio controls 2 and 3 with their receivers change after only a few milliseconds back to other frequencies. Using them, they can communicate undisturbed again. As shown above in the discussion of the four-lane highway in Figure 47, it is as if one of the boxes with its information were lost. In practice, however, the loss of information is very small due to the short dwell time at a certain frequency. Therefore, both receiver 2 and receiver 3 still receive (almost) the complete information. Thus, both are still able to drive their servos, motor controllers and other peripherals reliably.

Risks for parallel operation with many radio controls
However, in most manuals of 2.4 GHz radio control systems, the manufacturers indicate that the signal quality will be poor when many transmitters communicate in the same place at the same time with their receivers. In this case more and more frequency sequences occur in parallel, and so the probability of a frequency collision increases. So it is also shown in Figure 49 with the sequences 2 and 3. In most cases, however, it is also noted that it is probably rather the airspace, the lake or the race track that only allows the simultaneous operation of a certain number of models due to the limited size.

The ISM band is indeed not only allowed for the RC alone, as discussed at the beginning of this book. Also other communication systems such as wireless Internet use the spread spectrum. Again, the signal quality decreases in principle with a growing number of devices, because the random collisions, and thus the interferences, increase.

The comparison with whispering voices
In Section 5.3 the spread spectrum technology was compared with the whisper of human language. That a whisper corresponds to a

smaller power is also true in technology. The different languages German, English and Chinese correspond with the FHSS sequence 1, 2 and 3. At the binding the receiver must therefore learn the sequence. Binding would accordingly mean that the human being has to learn German, English and Chinese to understand the various whispering voices in the room.

5.5 Direct sequence spread spectrum (DSSS)

At the beginning of the explanations of a different technology, namely DSSS, the whispering voice will be used as a comparison. This is also spread spectrum technology, but the implementation is different than FHSS.

Figure 50 shows again the PSK modulation which was already discussed in Section 5.2. If one were to compare it with the signal from a whispering voice, one would undoubtedly find that it has a much more complicated structure than the PSK modulation which is shown here. It could even be seen that the whispering voice contains much more information.

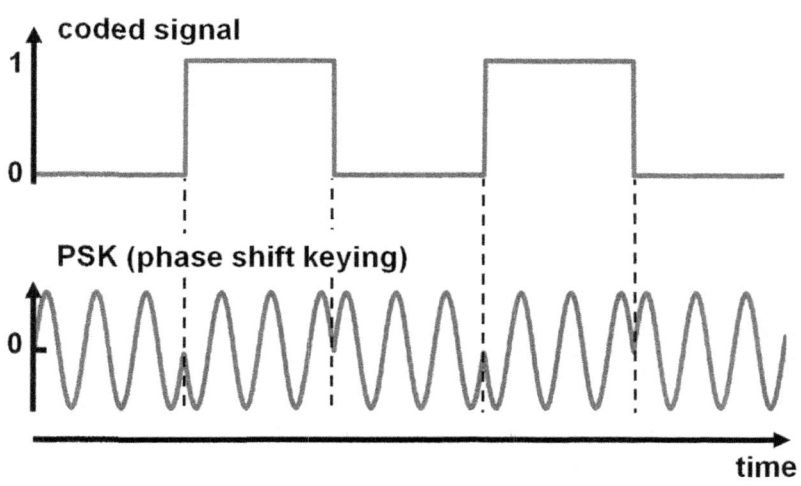

Figure 50: Coded signal and its PSK modulation

So with DSSS ways have been looked for to achieve a more complex signal structure. Figure 51 shows the coded signal as shown above, which is now linked with so-called pseudo-noise. Pseudo-noise means about the same as 'imitated noise'. It is, in practice, a defined sequence of zeros and ones. These change their value more frequently than the coded signal. The coded signal and pseudo-noise are now linked together with a so-called EXOR function. This means that the result is equal to 1 if one of the two signals is equal to 1 and the other is 0. If both signals are equal to 0 or equal to 1, the result is always 0.

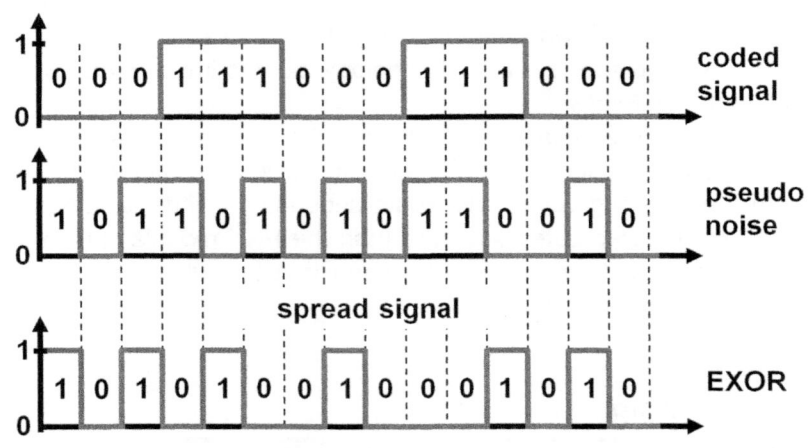

Figure 51: Preparation of the spread signal in the transmitter

At this point no signal theory will be done. However, if one were to compare the bandwidth of the top signal with the one at the bottom, it could be shown that the bottom one has a much larger bandwidth. Thus, the EXOR operation of the original signal with the pseudo-noise produces a spread signal. So this also corresponds to the spread spectrum technology.

In any case, one can already recognize in the example in the figure that the spread signal at the bottom has a more complex structure than the signal at the top and it is somewhat closer to the whispering voice. Again, there are many options for the high-frequency generation, and as an example PSK could also be used here.

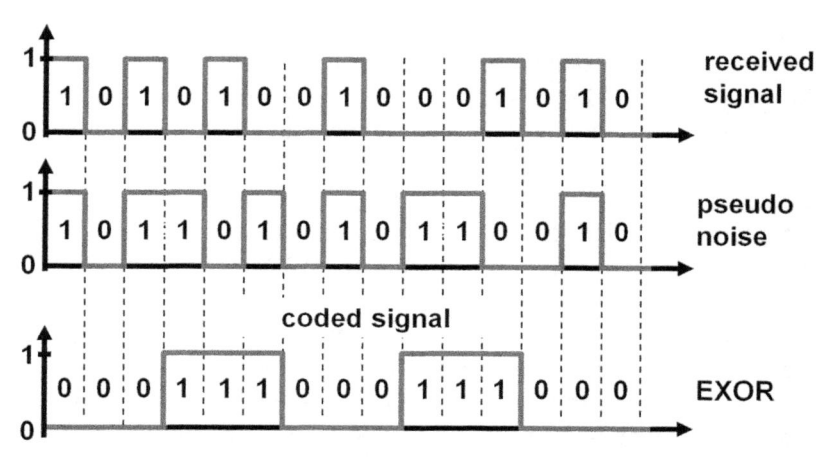

Figure 52: Recovery of the coded signal in the receiver

In the receiver, first a PSK demodulation is performed. To restore the original encoded signal, the receiver must also know the exact sequence of 1 and 0 of the pseudo-noise. So at binding with DSSS, the transmitter tells that sequence to the receiver. Direct sequence spread spectrum means here that the sequence will be directly linked to the signal. If the received signal is again EXOR-combined with the pseudo-noise, then the original coded signal is obtained again. This is illustrated in Figure 52. This can then be processed further again, so that the servos get the correct PWM signals.

In Figures 51 and 52, the sequence of 1 and 0 of the pseudo noise changes the value three times faster than that of the originally coded signal. It could be shown with signal theory that also the bandwidth of the spread signal increases by a factor of three in comparison to the coded signal. In practice, the value of the pseudo-noise changes much more frequently, which also leads to a much more complex signal structure. Then also the bandwidth of this spread signal and the information quantity are much higher.

5.6 Combination of FHSS and DSSS and other terms

Because FHSS and DSSS have in signal theory different advantages and disadvantages due to their characteristics, one can't definitively say which is better suited for radio controls in model construction. However, in Sections 5.4 and 5.5, they were treated in an idealized manner. In practice, each manufacturer integrates special properties such as a specific frequency sequence or the number of channels in FHSS or a special pseudo-noise sequence with DSSS.

Moreover, combinations of both are commonly used. Thus it is possible, for example, to spread the coded signal first with DSSS, but not quite as strong, and thereafter use FHSS. You could also go the other way by first using a FHSS and then a DSSS.

In diversity systems with two full transmission and reception units including two antennas each, it is also quite possible to use two different frequency sequences in FHSS or two different pseudo-noise sequences for DSSS. Thus, the transmission reliability is further improved.

In model construction, many manufacturers have defined their own terms for the 2.4 GHz technology. However, these are always based on either FHSS, DSSS or combinations. Examples are: AFHSS (Advanced FHSS), DMSS (Dual Modulation Spectrum System), DSM2 (Digital Spectrum Modulation 2), DSMX (Digital Spectrum Modulation X), FASST (Futaba Advanced Spread Spectrum Technology), FHSS-4T or S-FHSS.

Compatibility of the systems

With radio controls which operate in the MHz range and with PPM, you can combine the transmitters and receivers of different manufacturers in almost any way with each other. You just have to make sure that both are equipped with the same pair of crystals. The crystal with the label TX is designed for the transmitter and RX is for the receiver.

The different types of modulation in the 2.4 GHz band can be combined in many ways. Therefore, it is very often not possible to use different brands for transmitters and receivers together.

However, there are manufacturers who run their systems compatible with others. Thus at the time of purchase it is essential to pay attention to the compatibility of transmitters and receivers. But in most cases you will in any event use the transmitter and receiver from the same manufacturer.

Power measurement in dBm
Receive and transmit power are often expressed in units of dBm. This is a value in decibels, which refers to the relative power 1 mW. The power is calculated as

$$Power\ in\ mW = 1\ mW \cdot 10^{Value\ in\ dBm/10}$$

Example 4
The received signal of a radio control receiver will be measured with −30 dBm. Calculate the power in mW. It is: 1 mW x 10$^{-30/10}$ = 0.001 mW. If one wants to calculate the value in dBm from the value in mW, the logarithm is used (on the calculator, 'log'). The value in dBm = 10 x log (value in mW). 10 x log (0.001) = −30 dBm.

The received signal is always much weaker than the transmitted signal. As the wave propagates in spheroid circles from the center of the transmitting antenna, the receiving antenna can use less and less from the total radiated power with increasing distance. This was discussed in Section 3.2. As the example shows, decibel has a logarithmic scale. Table 2 shows the relation between the power in mW and dBm.

Power in mW	100	10	1	0.1	0.01	0.001 = 1 μW	0.0001
Power in dBm	20	10	0	−10	−20	−30	−40

Table 2: Power in mW or dBm

As the table shows, the received power decreases with increasing negative dBm values. A received power with −40 dBm is therefore smaller than one with −30 dBm.

2.4 GHz forever?

In the first years of the 21st century there was a dramatic increase in wireless communication devices at 2.4 GHz. This will lead sooner or later to the need for additional frequency bands. In the 1990s it was still much more complex to run a 2.4 GHz communication, due to technology. Today, however, even much higher frequencies can be realized for wireless signal transmission. And with technological advances, always higher frequency bands will be developed in the future. The needs of the modern communication society for more and faster wireless connections can only be solved this way.

The authorities will provide additional frequencies for the various stakeholders. It is quite possible that the RC for model construction will send on other frequency bands in the future. However, the technologies presented in this book and the properties of the GHz waves will remain. The behavior is indeed often compared with that of light waves. If the frequencies are even higher, this comparison is even more relevant. One can be curious about what is to come.

6. User interface and programming

The transmitters can be operated in various ways. The individual functions are controlled via the sticks, steering wheels, rotary switches or buttons. Especially with inexpensive modeling products, such as mini-helicopters, small aircraft or toy cars, complete sets are available which include a simple 2.4 GHz radio control. These are usually equipped with only a few functions, for example only with sticks and possibly with the possibility of trimming. The focus of this chapter is not aimed at these radio controls. It is rather aimed at the more expensive RCs, which are equipped with many other features. All these are programmable. They offer more options, such as being able to save the settings of a model, mixed functions, servo adjustment, or the so-called teacher-student operation. The following sections will explain the properties of these functions.

6.1 Stick, steering wheel, rotary encoders, switches and buttons

Figure 53 shows an image of part of a transmitter, where the stick, rotary switches, switches and buttons can be seen.
As in the past, the main functions are still controlled by the cross sticks. These have dominated the development of radio controls since the beginning. But there are only two cross sticks in standard radio controls, one for the left hand and one for the right hand. This allows the control of four channels or four functions. The cross sticks are always provided for the main functions. These are the rudder, elevator, ailerons and throttle in model airplanes. For model helicopters there are elevator, aileron, throttle/pitch and yaw. The exact channel assignment for these models is discussed further below. The main functions of car models are steering wheel, gas and brake; in model ships there are rudder and throttle. In both of these model types, the colt transmitters discussed above are often used.

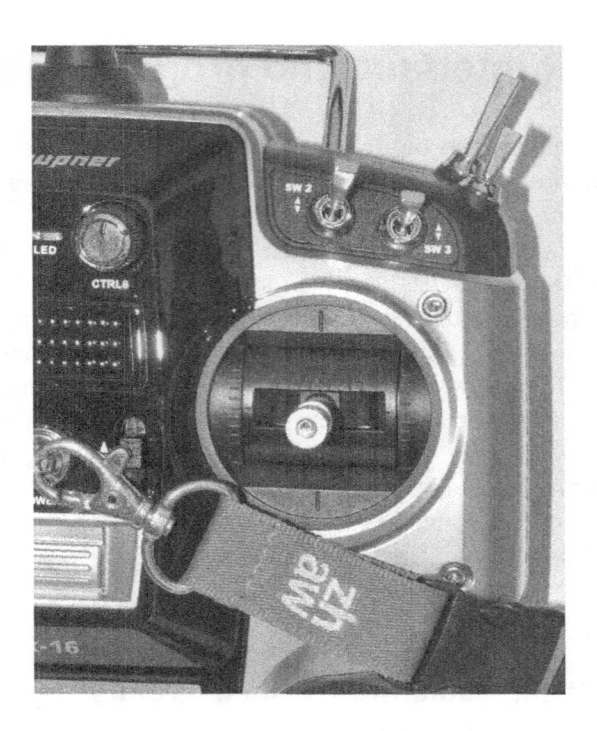

Figure 53: Stick, rotary encoders, switches and buttons

Rotary encoders

However, most of today's radio controls provide more channels; usually there are six, eight or twelve. In addition to the two cross sticks, several channels will therefore still be controlled with rotary encoders or switches.

The rotary encoders are usually realized with potentiometers. The servos will move according to the position of the encoder. Since during normal operation the hands have plenty to do with the two cross sticks, the rotary encoders are responsible for special functions. To use it, your hands have to release the stick for a short time. Functions for this purpose are for example the movement of flaps or the extension and retraction of landing gear with a servo. In ship models, the options are various; possibilities include the raising and lowering of a crane or the rotation of a light position.

Switches and buttons

If an eight- or twelve-channel radio control has two cross sticks and two encoders installed, two or six channels must still be controlled by additional switches or buttons. A switch can have two different positions or three if it has a central position. Many larger models have enough special features that are controlled. There are also retractable landing gears, which have an integrated electronics and mechanics, so which do not need to be controlled by a servo. With these, there are usually only two positions, 'extended landing gear' and 'retracted landing gear'. Further, jumps of parachutists from a model airplane can be triggered, colored vapors can be activated, model car lights can be switched on, ship model anchors can be raised, and so on. Of course, also with these functions the model pilot or captain must release the stick for a short time. Since only one switch must be turned over, however, this is often somewhat easier than the operation of a rotary encoder.

Switches and buttons do not only control functions, but they can also switch the effects of the transmitter itself. This means for example that they switch between things like model storage, operating and flight phases, mixers, dual rate or expo. These are described later in this chapter. As there are several manufacturers of RCs, so there are different philosophies concerning what the allocations of switches and buttons should be. There are RCs in which they are strongly assigned to either a function or to switch a transmitter effect. In others they can be freely assigned with the corresponding programming.

6.2 Programming

Today's RCs are all based on microcomputers. Microcomputers can be programmed for various functions. Strictly speaking, the model pilot or captain is not programming his radio control in the real sense. He rather configures it to his needs. However, since in the operating manuals 'programming' is always mentioned, this term shall be used here. Below it is shown what is necessary for a modern radio control transmitter so that it can be programmed.

Buttons and display

In Figure 54, buttons and displays are shown. They are required to perform the programming inputs, and to illustrate the effect of these. RCs are menu-driven. This means that you access the various functions via the user menus. These are discussed in the next sections.

However, the display has an important role, not only in programming. In normal transmission mode, there is plenty of further information which is important for the operation. Examples include the battery voltage of the transmitter or the current operating time of the model. Also, important data from the receiver can be shown via the telemetry. This will be discussed in Chapter 7. Examples include the voltage of the receiver battery or the strength of the received signal.

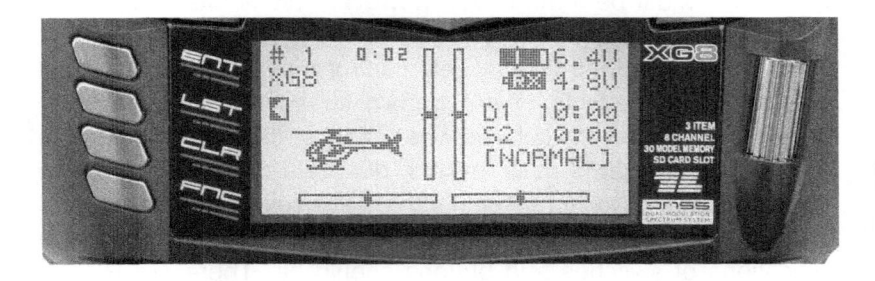

Figure 54: Display and buttons of a programmable transmitter

Acoustic signals

The feedback of the programming can also be done through acoustic signals. However, these are even more important in the normal control mode. Since the model pilot or captain then observes his model, he must devote full concentration to it, so it is usually not possible for him to look at the display. Therefore, the transmitters also have acoustic possibilities. These are often small loudspeakers, which are realized with piezoelectric elements. Thus, at least one warning will be issued with beeps when the battery voltage in the transmitter is too low.

With very sophisticated RCs, a speech output is included. This way, even more information can be transferred to the user, while he can still observe the model.

Model memory
The ambitious model pilot or captain usually has more than just one model. It is quite possible that he brings several of them to the race track, to the airfield or the lake. These may be so different that at least the switches and rotary encoders, but sometimes even the cross sticks, control completely different functions.

Therefore, the programmable radio controls have several model memories. These are configured and programmed for one model. In the displayed menu, you can switch between these models by buttons.

Operating phases
In some situations, it may be quite meaningful that the stick does not cause the same effect with a model every time. So it is possible that the servo movement for gentle flight must be larger than for a speed flight, as this has a much greater effect due to aerodynamic reasons. For a quadrocopter, it may be possible that during the flight one must switch between an angular control and an angular velocity control of the pitch and roll axis. For model helicopters there may be a phase of operation, 'auto-rotation'. In this position, the pitch position of the rotor blades is changed accordingly, and the mixer to the tail rotor is also set differently. With model aircraft, the operating phases are also called flight phases. With model cars it could even be necessary to have different servo settings for different speeds or terrain conditions. Even with ship models, the rudder movement may be varied.

There are several ways in which this may be implemented in the radio controls of the various manufacturers. In most cases it is possible to program a plurality of phases of operation in one model memory.

The switching of the operation phases is then made with a button or a switch. This is either permanently assigned to this switching function or it can be programmed to it.

Binding

What binding is from a technical point of view was explained in Chapter 5, in the explanation of FHSS and DSSS. Repeated briefly, the receiver must get to know the frequency sequence with FHSS and the pseudo-noise signal with DSSS. In the commonly used combinations of FHSS and DSSS it must get to know both. But the user is now also interested in how the binding is made in practice. There are various ways depending on the manufacturer. Two of them will be discussed below.

At the wired binding, the receiver is connected via a special cable to the transmitter. Then a corresponding command on the menu is started. Thus, the transmitter transfers the required data to the receiver.

Binding also works wirelessly with many systems. Also then it is triggered by a menu command. Additionally, it may also be necessary with some manufacturers that a button is pressed at the receiver. Then the transmitter and receiver communicate using 2.4 GHz signals, with a preset frequency sequence from the factory. That only serves to transmit the necessary data. Wireless binding is mainly used in systems with telemetry, since then the receiver can inform the transmitter whether it has received all data correctly or not.

After the binding was performed correctly, this is indicated either by the sender on the display or a status LED on the receiver. In some systems it also beeps, or a combination of the above.

PC interface

So that the radio control can be programmed by the user at all, special software is needed, so-called firmware. This provides all the functions and interfaces that are required for programming. The manufacturers expand the firmware from time to time. So it is quite possible that a radio control which is updated with new firmware features more options than at the time of purchase.

The new firmware must come in some way into the radio control. An update is performed with a PC interface. Today this is usually the USB interface. The new firmware can be downloaded to the PC from the download area of the manufacturer's website. Through a special menu, the user is instructed to connect the radio control to the PC via the interface. Then the firmware can be installed on the RC. This is done with non-volatile memory. Thus, the firmware is still available after switching off the system or removing the batteries.

However, it should also be noted at this point that the user generally does not have to make any firmware updates. One can assume that new updates have been extensively tested by the manufacturer. Nevertheless, they sometimes bring the risk with it that errors can creep in, or in extreme cases old functions work differently than before. He who is happy with his radio control and its capabilities can take the position that he does not want to install new firmware. Moreover, not every model builder is a hardcore computer geek.

Memory cards

The situation is similar with the memory cards. Like PC interfaces, they are also not absolutely necessary components of radio controls. But for many features of modern RCs, their use is at least a major advantage.

In the simplest case, with the memory cards it is possible to create backup copies of each model memory. Thus, in the case of data loss, not all the functions have to be re-programmed laboriously by hand, but can simply be copied back. Also for new RCs from the same manufacturer, the functions and model memory cards are often compatible. Thus, these memory cards can often be copied to a new radio control. But there is no guarantee that this works properly and in all respects.

As discussed in the next chapter, using the technology of telemetry it is now possible to transmit sensor data from the model to the transmitter and to store these there. This could be, for example, the voltage profile of the receiver battery, the quality of the received signal or height measurements in a model aircraft. Thus, analyses of the data can be performed afterwards. These data are then transferred either directly to the PC via the interface as described above or they can be stored on the memory card.

The same brands are used as memory cards as are used in cameras or USB sticks. These are so-called flash EPROM memories.

Button cell batteries in the transmitter

Some radio controls also contain button cell batteries which are usually lithium batteries. These are of course not designed for the operation of the transmitter. This is ensured by the transmitter battery discussed in Section 2.5 ('Energy supply'). Sometimes in RCs the date and time are also stored. Therefore this small battery ensures that they are not lost when the main battery is changed. This battery is usually designed so that it does its work reliably for several years.

The model memory and other user-programmed functions are often stored in an internal flash memory, which is based on the same technology as the memory cards discussed above. For most RCs their content is therefore retained, even when the main battery is changed and they do not contain any built-in button cell battery. But this is not solved the same way with all brands and therefore before changing the battery it is worth checking in the manual how that behaves exactly with the loss of data.

6.3 Channel assignment

With the colt transmitters for RC cars and fast RC boats, the channel assignment is almost intuitively given. This was already discussed in Chapter 2. However, even for model airplanes and helicopters, corresponding standard stick assignments have been established with the handheld transmitters. They are also called modes. Figure 55 shows the default stick modes for model airplanes. Which mode is used depends on the individual preferences of the pilot. When the stick is released, it is drawn to the center by springs. Only upon throttling do most pilots prefer a gridding. When they release the stick, it remains in the set position. A metal plate presses on a plastic grid. Depending on which mode you want to use, you have to convert the corresponding stick for throttling. Figure 56 shows a detailed view of this. This is provided as standard in most remote controls. The operating instructions describe exactly which handles are necessary for converting.

model airplanes

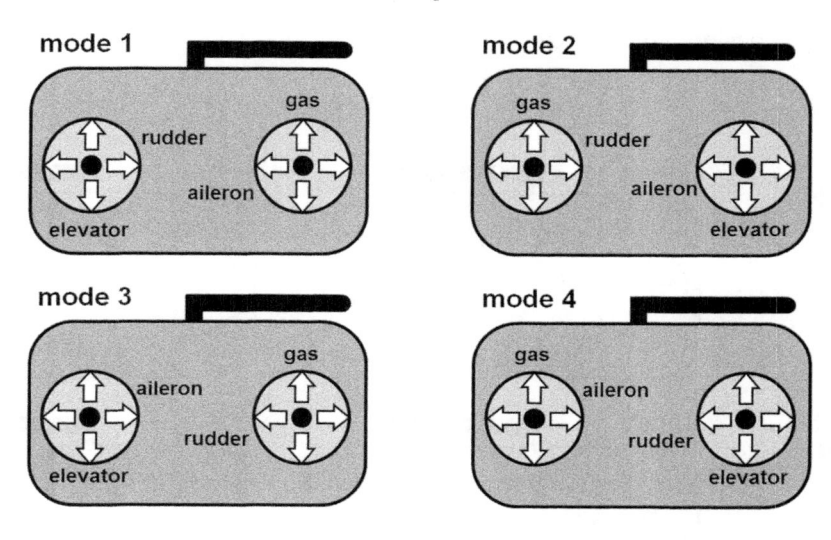

Figure 55: Standard modes for model airplanes

Figure 56: Gridding on a throttle stick

These modes also exist for model helicopters. A corresponding assignment is shown in Figure 57.

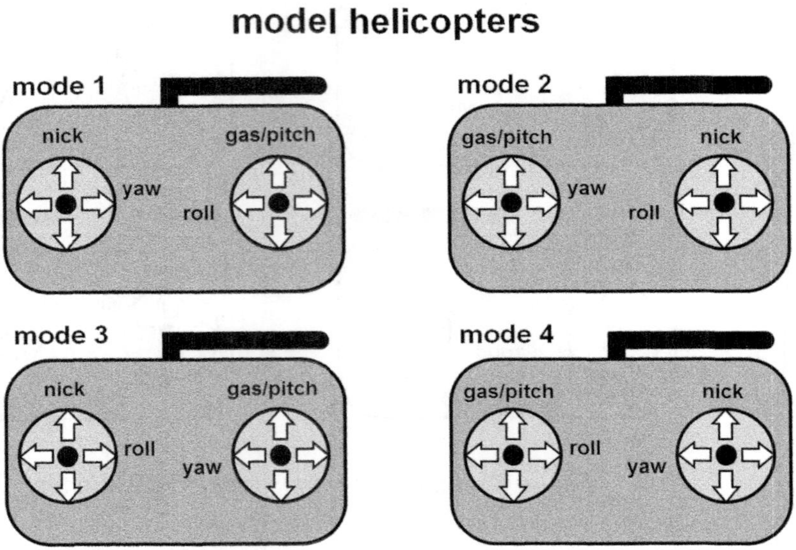

model helicopters

Figure 57: Standard modes for model helicopters

If a model memory is programmed, the mode must also be selected. The channels are then automatically assigned. The exact order depends on the manufacturer. The operating manuals will always state in which slot the servos must be plugged in for throttle, aileron, elevator and rudder and for pitch, roll or yaw. Alternatively, you can also try to find it out by yourself. You can push the stick of the corresponding function while a servo is plugged in a slot.

With car and ship models the sticks are not as consistently allocated to the individual functions. If it is assumed that the steering wheel or the rudder is like the rudder of an airplane model, you can of course also use the modes shown for the airplanes according to Figure 55.

6.4   Servo trim, reverse, travel adjustment

Figure 58 shows the travel range of a servo. Since there is nowadays a large diversity in all quality and price ranges, it can't be said conclusively for all exactly how big this is. The area which is used in the normal case approximately covers – depending on the servo – a range from 90° to 120°, that is, a quarter to one third of a revolution. This is illustrated in the figure by +/−100%. It has already been discussed above that servos have a position measurement. In some measurement principles, especially also with the one of a potentiometer, a mechanical stop is provided which limits the maximum possible travel.

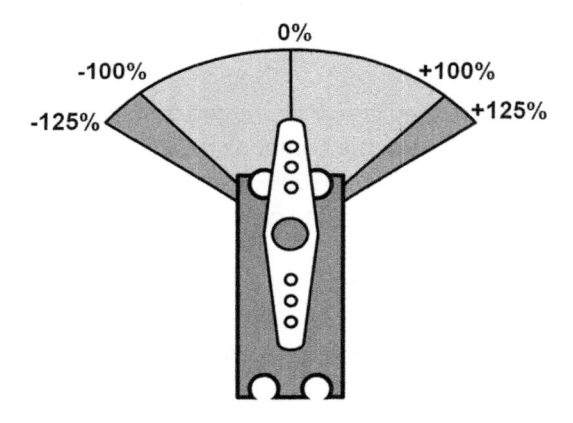

Figure 58: Travel range of a servo

How much travel range still remains over the +/−100% and up to its mechanical attack also depends on the servos. In systems with a non-contact position measurement, it is possible that the travel range can again be 50% or more in addition on both sides, or that no stop is found at all. It is also possible in some systems that although a mechanical attack is present, the servo electronics are prevented from being driven into this. But for the further considerations with regard to Figure 58, we assume that an additional 25% travel range on both sides is possible.

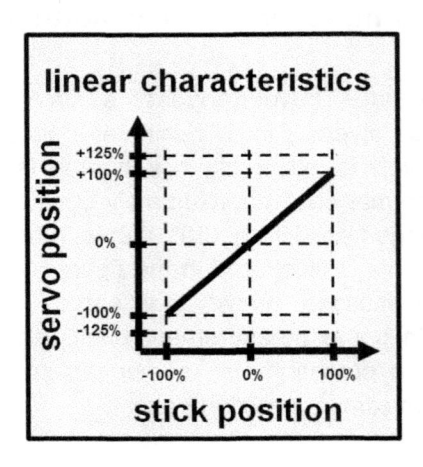

Figure 59: Linear characteristic

Figure 59 shows the so-called characteristic of the above-discussed servos. On the horizontal axis the position of the stick is shown. It can be varied from −100% (left) to +100% (right). Without any additional settings in the radio control, the characteristic is as shown. With a stick position of −100%, the servo travel is also −100%, and with a stick position of +100%, the servo travel is also +100%. In other words, when the stick is to the far left, the horn of the servo is also left, and when the stick is to the right, the horn of the servo is also right. This is a so-called linear characteristic, so the line is straight. Thus, an increase in throttle position corresponds to an equal increase in the amount of travel.

Reverse

In many cases it happens that you notice during the first function test after assembly that the servo moves in exactly the opposite directions as desired. In this case, the travel range can be inverted with the reverse function. When the stick is in the left position, the servo moves to the right, and if the stick is on the right, it moves to the left. This function is found in almost all radio control transmitters. It can either be programmed in the menu or set directly by switches with simpler transmitters.

Trim

The possibility of the trim is also provided by all radio control transmitters. The effect of the servos on the rudder of the aircraft or ship model or on the steering axis of the car model can be adjusted during installation to the best of our knowledge. However, only actual use will show if the model flies, sails or drives straight ahead. If this is not the case, trimming is required.

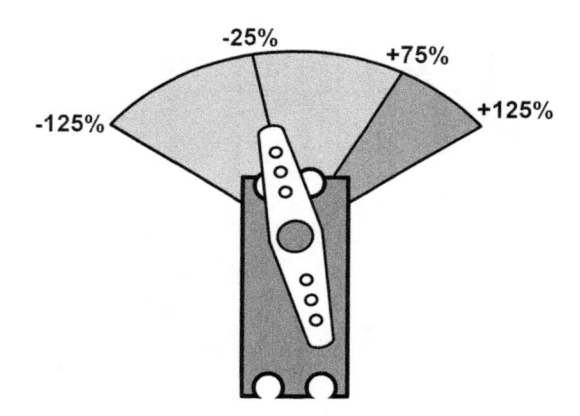

Figure 60: Travel range of a servo with negative trim

Figure 60 shows what this means for a negative trim, so a left trim. In the neutral position of the stick the servo horn is rotated slightly to the left. Compared with Figure 58, the two extreme positions are then moved to the left. Of course, the same applies correspondingly to the right side for a positive trim. It must be noted that the servo will then rotate closer to the mechanical stop on one side. In the illustration in Figure 60, even the maximum possible rotation would be reached with a trim of −25%. Normally, possible mechanical limits are not yet reached at the maximum trim. However, one should always test this. Namely, if the servo moves into a limit, it is often not noticed, either by the transmitter or by the receiver, because a corresponding feedback may be absent. The result is then a noise that sounds like growling. Servos can also growl without their position being at the mechanical limit. Then it should be checked whether the rudder or the rod is moving too stiffly. Growl means, most often, that the servo is overloaded in some way. This is also accompanied by an increase in current and should be avoided if possible. Trimming is realized in most radio controls with an additional lever near the stick. Figure 61 shows a possible implementation. With a cross stick with two functions, there are of course two such levers.

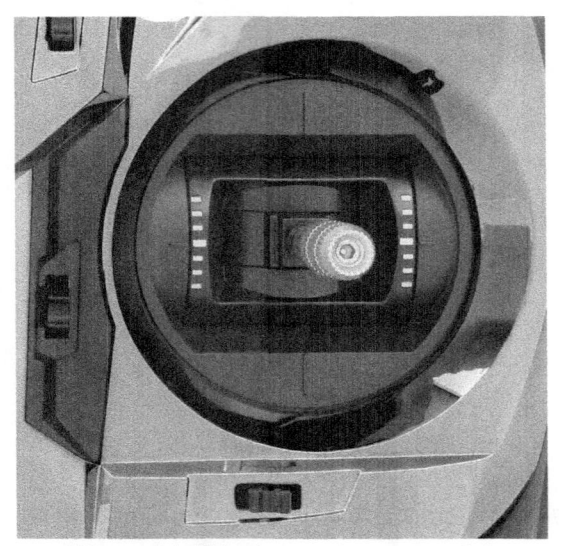

Figure 61: Trim lever at a cross stick

To trim servos in which the center position is not exactly where you would actually expect it, there is with some RCs also the option to adjust only this center slightly. However, that is usually not undertaken via a lever, but is set in the programming menu. In the manuals it is termed 'sub-trim' or similar.

6.5 Dual rate, expo and further functions

In Figure 59 a so-called linear characteristic of the servo was shown. It is used in most applications in model construction. Moreover, it is always the default. But there are also other types of curves. Their use for the one or the other special case is quite reasonable.

Dual rate

Above several possible operating phases of a model were referred to. If we take as an example a model airplane with which you want to fly gently on one occasion and another time you want to operate speed flight, it could be useful for aerodynamic reasons that the stick does not have the same effect on the model in both cases. A certain rudder angle has less influence at low speeds than in speed flight. This can be solved with dual rate. Figures 62, 63 and 64 show curves with dual rate 100%, dual rate 50% and dual rate 125%, respectively.

Dual rate 100% in Figure 62 is the same characteristic as already shown in Figure 59. So this is a normal linear characteristic. In the discussion of the next curve in Figure 63, a stick position of −100% only has a rudder of −50%, and a stick position of +100% corresponds to a rudder of +50%. With full stick the servo only moves half way compared to Figure 62. In the operating phase 'slow flight' one could use dual rate 100% and in the operating phase 'speed flight' one could use a programmed dual rate 50%. This means that the model can be controlled more sensitively with the same stick travel. The disadvantage of this is that you then only have half the servo travel available in this mode. Of course, you can realize this in the other direction as well.

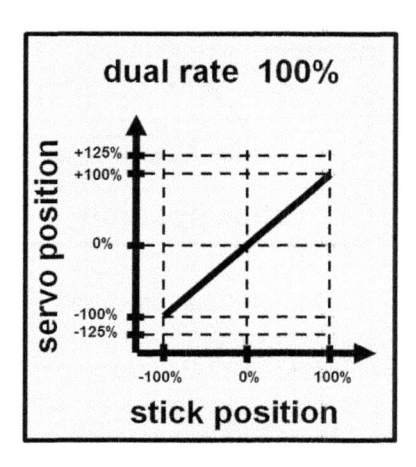

Figure 62: Dual rate 100%, characteristic curve corresponds to standard linear

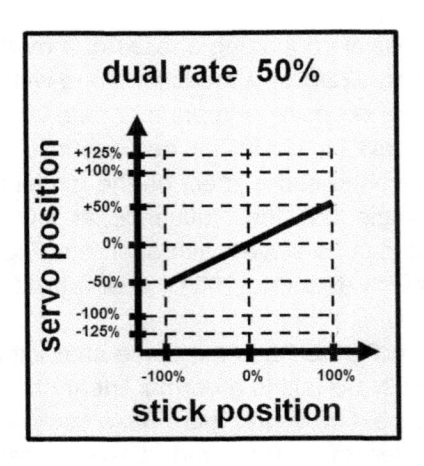

Figure 63: Dual rate 50%

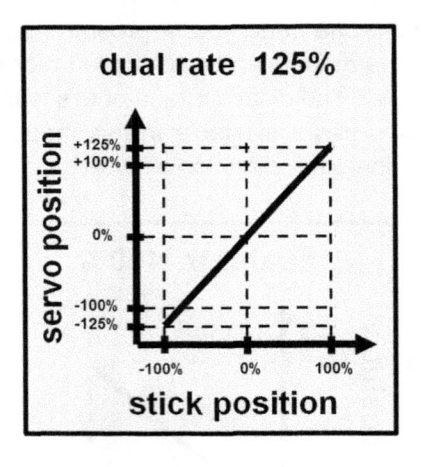

Figure 64: Dual rate 125%

Figure 64 shows a dual rate of 125%. The servo now rotates at the maximum positions of the stick more than with 100% dual rate. Here it would be for example the entire travel range from −125% to +125% of Figure 58. As also mentioned in the section on the trim, it must always be ensured that the servo does not run into the mechanical limitations.

Expo function

Actually the dual rate characteristic is linear, so they are all represented by straight lines. It might now be, however, that the model pilot or captain would like to have for small stick movements only very small deflections at the servo and that they will increase disproportionately for larger stick movements. He then has on the one hand the advantage that he can control the model very sensitively around the center position of the stick. On the other hand, he nevertheless has in the larger stick movements the possibility to reach full servo deflection. This function is actually called the 'exponential function'. In model construction, however, the expression 'expo' has established itself. It is therefore used here. Figure 65 shows an expo with 100%, and Figure 66 shows a slightly weaker variant with expo 50%. The percentage values are chosen only qualitatively in most RCs and simply give an indication of how strongly the characteristic differs from the linear characteristic shown in Figure 59. The effect on the model must always be tested in practice anyway.

Everything is also available with some RCs with negative percentage values. Then small stick movements have very large effects on the servo travel. Figure 67 shows such a case. But larger stick movements then no longer have such a great effect.

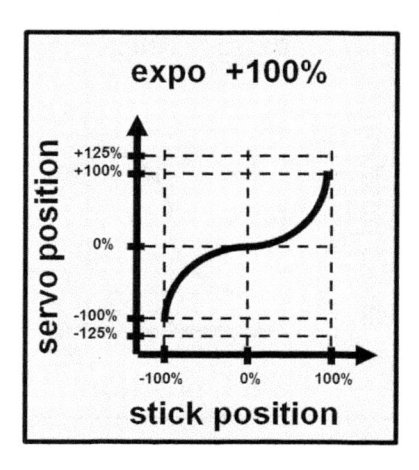

Figure 65: Expo 100%

103

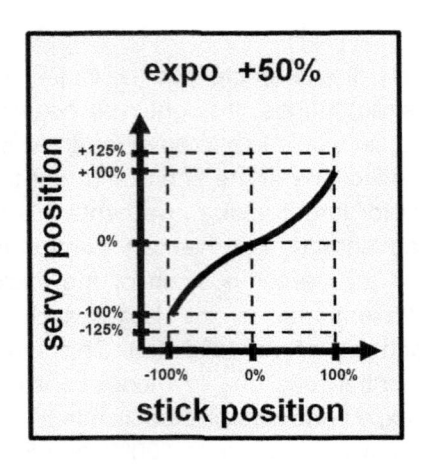

Figure 66: Expo 50%

Finally it should be mentioned that some RCs also allow combinations of dual rate and expo. In the example that is shown in Figure 68, an expo of 100% is combined with a dual rate of 50%. Some model builders may see this as pure gimmick, but the author has met hobbyists who have programmed such servo characteristics for special applications.

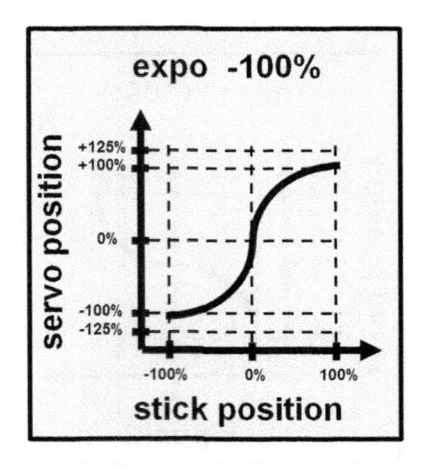

Figure 67: Expo −100%

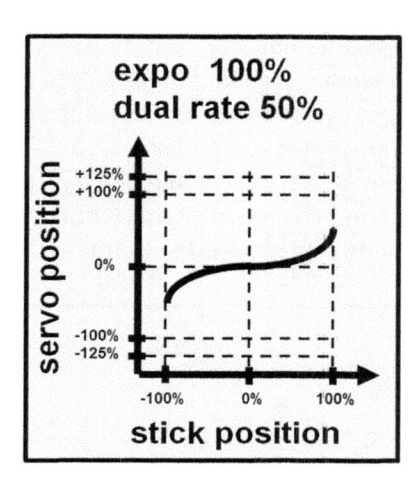

Figure 68: Combination of expo and dual rate

Further functions

Many RC transmitters offer the possibility to define characteristic curves with interpolation values. For example, in the characteristic shown in Figure 69, three interpolation values are necessary: at −100% stick, the servo travel is −50%, at 0% stick – i.e. the center position – the servo travel is 0% and at +100% stick the servo travel is also +100%. Between them the values are interpolated and that then gives the characteristic curve with a bend in the middle. Such a characteristic can for example be used as a gas characteristic for a combustion engine. When the throttle stick is with −100% at the very bottom, then the engine should run at idle. That would be a servo travel of −50% for this example. Next you might want at the center position, so at 0% throttle position, a forward flight or a forward motion with preferred pace. Then the servo position is in this example at 0%. For the maximum of 100% stick position you want full throttle, which corresponds in the example to +100% servo travel.

Figure 70 shows another self-defined characteristic, which could correspond, for example, to the motor characteristic of coaxial helicopters or quadrocopters. Here the 0% stick position, the center position, was thus defined specifically with the servo travel that corresponds to the hovering flight. As a further point, the maximum and minimum values are defined. It should also be mentioned at this point that this is actually not the servo travel, but the control of the motor controller. Coaxial helicopters and quadrocopters are indeed mostly controlled by the motor speed. All characteristics shown in this chapter apply not only to the control of servos, but to all kinds of peripheral systems

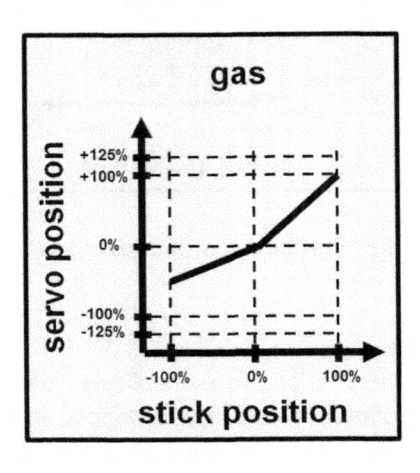

Figure 69: Idle gas

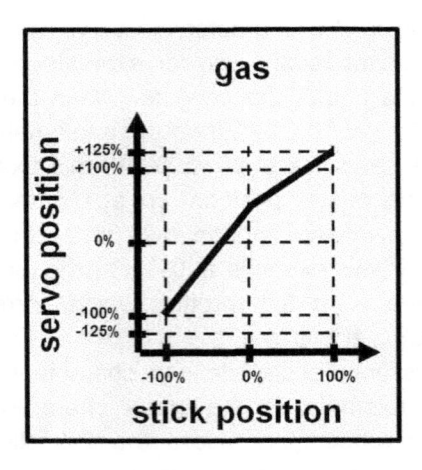

Figure 70: Hovering flight

106

6.6 Mixer

Up to this point only the characteristic curves of single servos were changed. It acts thereby in each case exactly one channel on a single servo or motor controller. However, in practice it is very often the case that for example two channels together have to operate on two servos. Hence there is a coupling between them. Figure 71 illustrates such a case symbolically. This coupling is called 'mixer' in model construction. Again, there are different possibilities how this is implemented in the various radio control transmitters. Some manufacturers provide only some basic functions such as the so-called V-tail. In others there are for almost any case some preconfigured functions for mixing which can be freely combined. In the following some typical cases of such mixing functions are discussed, right across model construction. There is no claim to completeness.

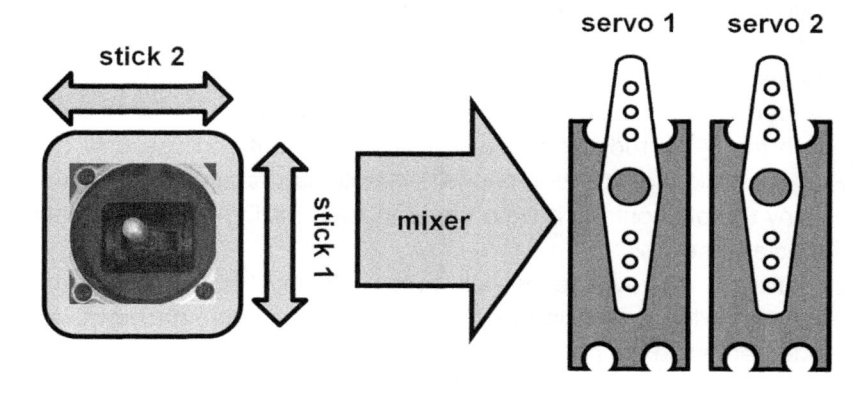

Figure 71: Two sticks are mixed on two servos

One channel operates on two or more servos

This is the most commonly occurring case. For large car models, a single servo often does not have a sufficiently large torque to perform the task alone. Therefore, an additional servo needs to be mixed, which is plugged into a different slot on the receiver and which performs the same function. There are many more examples. This is also the case with ship models with two arranged rudders on the left and the right, which are controlled by separate servos. The same applies when two servos drive the left and right landing flap of a model aircraft.

For model helicopters it is certainly worth mentioning the coupling of the pitch and the throttle function. If the rotor blades are pitched more strongly, more gas is necessary to keep the rotor speed.

It may also happen that two servos must move opposite to each other, for example in the control of the ailerons with larger model aircraft. It may also occur that the downward deflection should be less than that of the upward. This follows basic aerodynamic laws, according to which the downward deflected rudder has a larger effect. With evenly deflected rudders, the result is a deflection of the aircraft from the desired direction. In this case, these aileron servos must be programmed similarly to Figure 69 with two self-defined characteristics, and they have to be mixed on the same channel.

Two channels operate on two or more servos

If an aircraft has a V-tail, the function of the elevator and of the rudder is combined. Thus the two channels of the elevator and rudder must be mixed together. If you use only the elevator stick, both the V-tail rudders move evenly, whereas if you use only the rudder stick, they move in opposite directions. As discussed above, many remote controls have a preconfigured mixing function for this special case.

There are, however, especially in aircraft and helicopter models, further mixing functions in which multiple channels can be mixed together. An example is the combination of ailerons and rudders for a nice curve flight, which is only active when the aileron function is operating. Upon actuation of the rudder function, this is working alone.

With car models there is the combination or mixing of gas and brake. If only the brake is actuated, this servo acts alone. However, if the throttle is withdrawn, depending on the model it can also be mixed to have an additional effect on the brake servo.

These few examples show that the mixing functions are indispensable for modern model construction, so they make up a significant part of the configuration. This book does not go deep into detail in any area of model construction. So regarding the mixer, the interested reader is referred to the relevant specialist literature of model aircraft, helicopters, cars and boats.

6.7 Teacher–student operation

In order that an inexperienced model pilot or captain can be supported by a more experienced colleague, many RCs have a teacher–student function. Within this context, the experienced pilot or captain – the teacher – and the inexperienced pilot or captain – the student – each have their own RC transmitter in their hand. But they both control the same model. A typical situation is, using the example of a model airplane, as follows. The teacher starts the model and brings it to a reasonable altitude. In this phase of the flight the student may now take control. If he inadvertently brings the model into a critical attitude, the teacher immediately takes back control of the aircraft. Depending on the experience gained, the student can gradually control all phases of flight from takeoff to landing.

Connection of transmitters via teacher–student cable
In most cases, the two transmitters are connected via a special cable. This is relatively short, maybe about two meters long. This also makes sense because the two model pilots or captains are standing next to each other anyway, since they also discuss together who has taken control of the model.

Thus, for most systems only the high-frequency amplifier of one of the transmitters sends its signals to the receiver. This is usually that of the teacher. He is also the one who determines with a switch or button whether the student can control or whether he takes back the control.

For compatibility reasons, the connection of the transmitters works best for radio controls of the same manufacturer. In more complicated models, the model memory and programming of the most important functions must be equal or at least similar. Also the setting of the trim should be the same. It simplifies things considerably if both RCs work the same way. In addition, the manufacturers also often have their own systems in the connecting cables and connectors.

If only the high-frequency amplifier of one transmitter communicates with the receiver, then the other may also be a MHz system. As was shown in the last chapter, today's transmission signals are mostly not compatible with each other. But the manufacturers are striving in the teacher–student operation to achieve compatibility at least in their own systems (MHz or GHz channels).

But even within the one brand, PPM is often the only compatible signal. Thus, it is often used for communication via the connection cable. If you want to connect a teacher transmitter to a student transmitter from a different manufacturer, then there is usually only a small chance that both can communicate with each other via PPM. In this case, one must know the pin assignments, i.e. on which pin which signal is present. You will often find this information online. Then you have to get the proper connectors and solder a cable by yourself.

Connecting of transmitters via telemetry
Today's 2.4 GHz RCs are often telemetry-capable. This is explained in more detail in Chapter 7. In this case, each transmitter also has a built-in receiver through which it can receive data. This is also called 'return channel'. If one now has two 2.4 GHz transmitters from the same manufacturer, both of which are telemetry-capable, then the connection cable for the teacher–student function can also be omitted, if the software allows it. In this case, the two transmitters are able to communicate via the return channel. Whether then only one transmitter communicates with the receiver of the model and the other transmitter or whether both transmitters communicate with the model is dependent on the individual manufacturers. However, since compatibility is limited to the brand and it is also necessary that a return channel is available, wireless teacher–student operation is quite rarely used.

In the teacher–student operation, two transmitters always work together with one receiver. Therefore, this configuration is more complicated to use than if only one transmitter were communicating with one receiver. An important principle of the teacher–student operation should therefore be at the end of the chapter and always be heeded:

In the teacher–student operation all functions should be carefully checked before starting the model. This concerns in particular the transfer of commands from teacher to student and vice versa.

6.8 Programming a microprocessor at the receiver

The 'hold' and 'fail safe' functions were presented in Section 2.2 and in the discussion of PCM in Figure 44. They are also most often programmed via the transmitter which then transmits the signals to the receiver. The receiver can then be programmed to only a very limited extent, apart from this one exception. The entire configuration of models and everything that has been discussed above is thus undertaken in the transmitter.

And yet there are also applications in modeling in which the programming is mostly undertaken in the model. Figure 72 shows a microprocessor system which is connected between the receiver and the servos

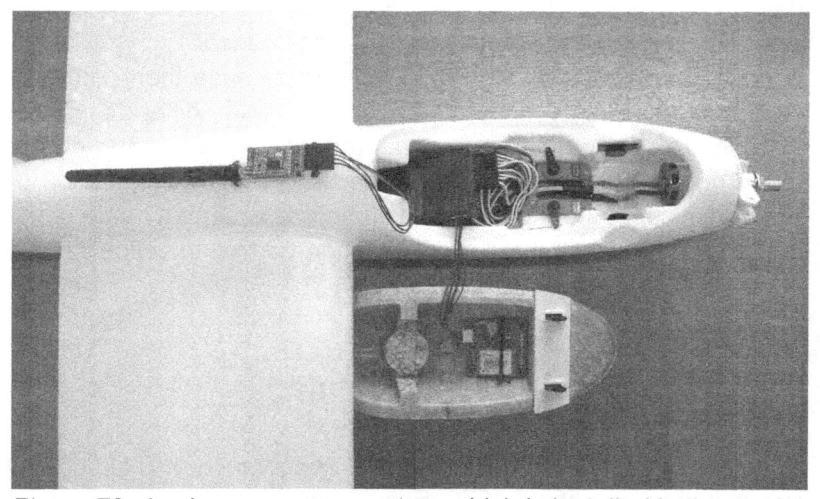

Figure 72: A microprocessor system which is installed in the model

Such microprocessor systems often include various sensors which contain additional information for direct processing in the model. This includes GPS, position sensors on all axes (which are also called Inertial Measurement Unit), air pressure sensors and more. Many model builders are excited about these technical possibilities. They program the microprocessor cards so that their planes, ships, cars, helicopters or quadrocopters can drive or fly autonomously to predefined points or routes. Here technical model construction merges with applied science. Some systems also include connections for video cameras. You can then store image data or send it with its own transmission to a ground station. Then the action from the perspective of the model can be followed directly with a monitor.

The system shown in the figure has a built-in GPS for the flying of routes and a transmitter mounted on top of the fuselage to transmit data to the ground station.

Of course, all functions discussed in Chapter 6 can also be programmed on these microprocessor cards. There are also quite a few manufacturers who provide routines for various functions. Often the codes are so-called open source codes, made available for free online for programmers.

The programming of such microprocessor boards is more complex than if undertaken by a preconfigured radio control transmitter. It is then carried out on the model side. However, on the transmitter side, only a few or no special functions are needed.

Actually, that almost makes more sense, because then you don't configure the radio control transmitter with different model memories for each model, but rather the model itself. However, it is today not foreseeable that the philosophy of the configuration or programming on the model side will become accepted in the broad model sports. To make a consideration in this direction should however be allowed at this point.

6.9   Radio controls in the future

The cross stick has characterized the appearance of radio control transmitters since about 1970. Even today, it makes sense that up to four functions can be controlled in this way with both hands. In the time since then, much has changed in terms of transmission, modulation and programming flexibility. But this main user interface has remained the same.

And yet today there are newer approaches how a pilot or captain can control his model. Figure 73 shows a smartphone which is used to control a quadrocopter.

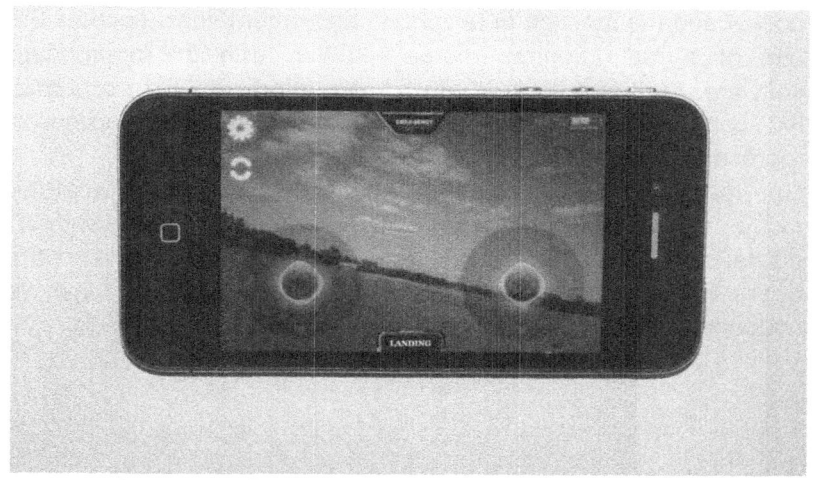

Figure 73: Smartphone to control a quadrocopter

Smartphones today are equipped with various sensors which detect the position and the orientation in all spatial angles. If you press a virtual button presented on the display, then this orientation is transmitted as a target value for the solid angles of the quadrocopter. So if you tilt the smartphone, the quadrocopter also tilts in the same direction. With a built-in compass in addition, the quadrocopter is able to measure its orientation compared to the pilot. So you can now switch to another mode in which the quadrocopter always tilts in the same direction as the smartphone, regardless of its relative position. This all works without any cross stick.

Front and rear built-in cameras can be transferred directly to the display. The image data can be stored on the smartphone itself. With an augmented reality known from modern computer science, even more information such as the height or virtual barriers can also be displayed on the same screen. With the operation of another virtual button the quadrocopter also loops autonomously.

Even at the beginning of the 21st century probably nobody would have been able to imagine that such devices would be available within a few years. Increasingly sophisticated sensors in combination with touch screens together with suitable software solve various tasks of everyday life. Everyone has such a device in their pocket and the function of telephone and other things, such as the one of a radio control, merge together with the appropriate software. Furthermore, it is much more interactive than a normal RC can be. The control of the model due to the inclination represents a totally new kind of user interface.

One may pause for a moment at this point and wonder what the future might bring in this regard. Will there be smartphones with 3D display and 3D interaction? It is quite possible that the radio control of the future no longer has any obvious similarities with what we understand and know by this term today.

7. Telemetry

Originally, radio controls were actually designed only to transmit control commands to the model. But of course it may also be interesting to transmit data from the model to the user. Figure 74 shows how this is meant.

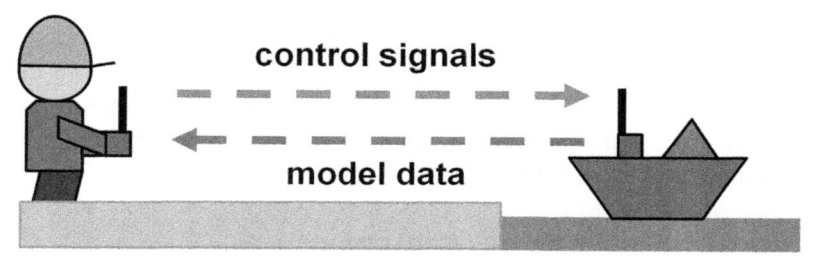

Figure 74: Telemetry

These data can be of quite different kinds. Examples are: strength and quality of the received signal, receiver battery voltage, motor current, motor speed, motor temperature, temperature of brushless motor controller, voltage of the drive battery, fuel tank capacity, height measurement, position measurement and speed. This all falls under the generic term 'telemetry', or 'bidirectional communication'.

The manufacturers of contemporary radio control systems provide telemetry-capable systems in different variants. Telemetry is already one of the basic features of modern transmitters and receivers. This chapter describes what kinds of bidirectional communication are applied in practice, how the data is made available to the user, what the different philosophies of the wiring there are and also what sensor principles are used.

7.1 Bidirectional communication

Independent data transfer

The simplest variant of bidirectional communication is data transmission which is independent of the RC. Thus the radio control does not have to be capable of telemetry but can concentrate fully on the communication from transmitter to receiver. An example of an independent data transmission is a video camera which sends the data directly to a PC or a smartphone. There, the data can either be stored directly or watched by a second user. Some manufacturers also offer special displays. These can be mounted directly on the remote control. The model pilot or captain can then easily watch them. When these systems also send at 2.4 GHz, some caution should be exercised as this may interfere with the radio control. This is especially the case if it sends with a high power. A bidirectional transmission that can transmit and receive data simultaneously with two independent channels in both directions is also called full duplex.

Data transmission on the return channel
Telemetry-capable radio control systems use one and the same system for two-way communication. Since the receiver antennas can also send with the appropriate electronics and the transmitter antennas can also receive, they can even be briefly switched into this mode. This is also known as data transmission on the return channel. The power on the return channel is usually less than that of the transmitter. In some systems, however, it can be set to an allowable maximum. It is also sent less frequently on the return channel.
The advantage of data transfer on the return channel in contrast to the independent data transfer is that data can never be sent and received in both directions simultaneously. In communications technology, this mode is also called half duplex. This type of transmission will never interfere with the control of the model. In addition, it also ensures that important information about the receiving system itself can be communicated. This includes information on the strength and quality of the received signal or the charging state of the receiver battery.

7.2 System overview and cabling

In the same way that most manufacturers follow their own standards for transmission, so the same applies with telemetry. In addition, the wiring of the sensors is not uniform. This section shows some system variations for data transmission via the return channel.

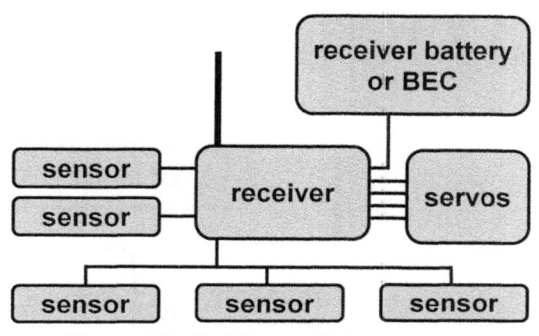

Figure 75: Cabling with bus system

In the illustration shown in Figure 75, the sensors are connected to a bus system. In some systems this is undertaken so that there is only one socket at the receiver. It is also possible for some systems to plug more sensors either directly to the receiver or to expand the sockets using a so-called Y-cable. 'Y' means in this case that two sockets lead to one plug.

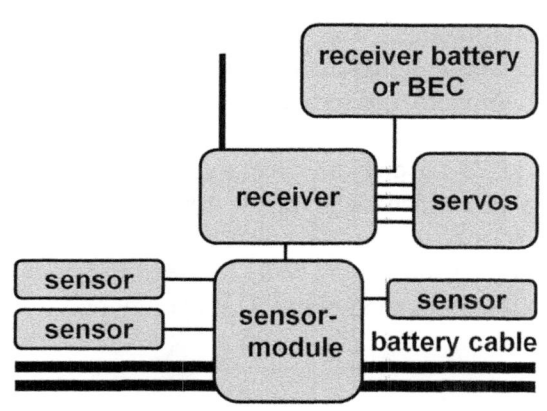

Figure 76: Cabling with sensor module

Figure 76 shows a further possibility for wiring with a specific sensor module. This is connected directly to the receiver. On the other hand, it also provides some slots for different sensors. In the version shown in the figure, the battery cables are also routed via the module. Thus, a current and voltage measurement is possible. The intelligent sensor technology can calculate the amount of charge in mAh, so the model pilot or captain even has information on the charge status of the battery.

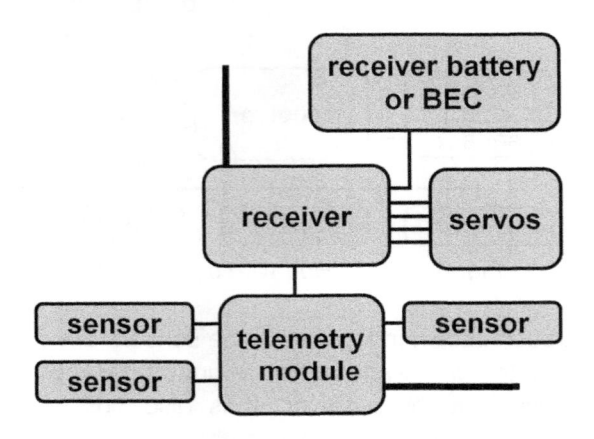

Figure 77: Solution with a telemetry module

Some systems do not use the receiver itself for the return channel, but an optional module. This will also ensure that information is sent and received in a coordinated manner in both directions. Also here the appropriate sensors can be plugged directly into the module. Figure 77 shows this solution. The author has tried with Figures 75 to 77 to give a summary of the available systems. There are also combinations of the three systems possible, for example a plurality of sensors could be connected in series on a bus, and a sensor module as well. The technology of telemetry is still young at the beginning of the 21st century and a standard has not yet emerged.

7.3 Visualization and perception of data

The much resulting data should of course also find a way to the model pilot or captain. However, because he must devote full concentration to the control of his model, the visualization of the data is not an easy issue to solve. Of course, it always depends on whether there is an assistant who can be devoted to data analysis. Some examples are discussed below.

Displays
In the telemetry-capable RC, the data are mostly presented on the display. Of course it is also possible to select the displayed sensor values in the menu. However, during the control of the model, the integrated display is often not in the pilot's field of view. Its main task is actually programming the radio control. For this reason, extra-large displays are often mounted on the upper edge of the transmitter. These are then more in the field of view and also especially designed for the display of telemetry data. Figure 78 shows such a solution.

Figure 78: Display for the visualization of telemetry data

Offline visualization
Since many model pilots and captains don't want to see any telemetry data during the control of their model, there is also the possibility of saving them. It is discussed in Section 6.2 that

119

modern RCs often have memory cards. In addition to the model configuration, the telemetry data can also be saved. If the previously stored data are viewed after model operation on the display or a PC, it is called offline visualization. But actually you do not necessarily need telemetry for that. There are also small flight recorder modules available which temporarily store the sensor data directly in the model. Even small cameras with built-in memory are actually flight recorders.

Voice output

Because the view must be directed to the model during control, manufacturers are looking at ways to use the other human senses. Modern microcomputer technology makes possible the development of radio controls which also provide voice information. Such systems are then also programmable in a way that, for example, via a push button operation only the information that is interesting to the user is communicated. This may include the voltages of the receiver or drive battery or the level of the received signal.

What data are useful?

'What should I do with all these data?', a hardcore model builder might ask. On the subject of telemetry, many model pilots and captains simply want to control their model, and to dedicate full concentration to this. Not everyone feels comfortable when a lot of information wants to be captured at the same time with different senses. One person might be disturbed by any data output while controlling the model. Another might want to know some important basic data, while yet another can't find enough information that he wants to monitor.

For that reason, everyone must answer the above question individually. It's important that everyone chooses the right path for himself and decides for himself what gives him added value in the control of the model. There are also various objectives that can be pursued with these telemetry data. This could be the comparison of power consumption for different propellers or the recording of the traveled distance using GPS. Your imagination has almost no limits here.

7.4 Sensors

Most systems today offer some basic types of sensors. The offer, however, is constantly being expanded. The specific manufacturer websites will always be more up to date than this book can be. However, this chapter will give an overview of the used sensors and how they work.

Voltage measurement

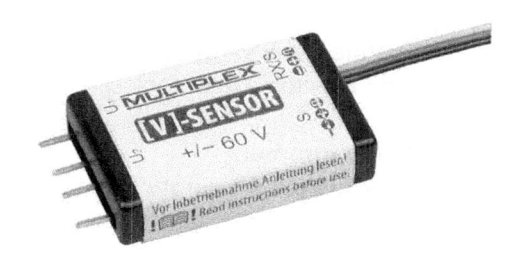

Figure 79: Voltage measurement

The voltage is mostly measured from the receiver battery or the drive battery. Figure 79 shows such a sensor. With this type, two voltages can be measured separately. For this, the positive and negative poles are connected on the corresponding inputs. The voltage measurement is relatively easy to realize. The applied voltage is converted with an analog-to-digital converter. Then a microprocessor brings the data into a form which can be processed further.

Current measurement
The current is measured in the simplest way, with a so-called shunt resistor. For this, the supply line must also be led via this sensor. Schematically, one can imagine something as is shown in Figure 76. The shunt resistor has a very small resistance of a few milliohms, and this results in a voltage drop which according to Ohm's law is equal to the resistance multiplied by the current. In

fact this generates small losses, because then there is less voltage available at the motor. However, these are comparably small.

Current-carrying conductors also cause magnetic fields. There is a proportional relation between current and magnetic field. Slightly more expensive current measurement devices use Hall sensors instead of shunt resistors. They can measure the magnetic field and thus the current without any loss.

Temperature measurement

For batteries, motors and controllers, a temperature measurement is especially useful when they are installed inside the fuselage and with no air flow. Platinum resistors are usually used for this purpose. These change their value as a function of temperature and can be evaluated by simple electronics.

Rotational speed measurement

Most rotational speed sensors operate optically. In this case, an infrared light diode emits a light beam, which is reflected from the rotating object, such as the propeller. A phototransistor evaluates the frequency of the returning light pulse and a microprocessor calculates the rotational speed. In addition, there is again the variant with Hall sensors. These measure the rotational speed more reliably. However, a magnetic plate must then be mounted on the rotating object.

GPS

Figure 80: GPS module

GPS modules are available for many applications. Figure 80 shows an example for model construction. These modules provide the user with three-dimensional position data. In addition to the coordinates, the height is also given. Since GPS has an accuracy of only about 10 m, the height data are usually not suitable for model construction as differences in the height of 10 meters are often crucial. For altitude measurement an air pressure sensor is therefore usually used.

Speed measurement with dynamic pressure
GPS modules measure the absolute speed using the reference of the ground. In aviation, that is also called ground speed. But flight models move in the medium air, so it may also be interesting to determine the so-called airspeed, or the speed relative to the air. Only with no wind are ground speed and airspeed identical; otherwise they are different.
In model construction telemetry sensors are also available for this. The airspeed is measured in most sensors by dynamic pressure sensors. For this purpose a sensor is placed at the front of the model. With increased speed the sensor measures a higher pressure. However, since the absolute pressure is dependent on the weather, a second sensor is mounted, protected so that it only measures the air pressure. The difference between these two sensor values is then a measure for the airspeed.

Altimeter
With increasing height above sea level, the air pressure decreases. Therefore many altimeters determine their values on this variable. A reference pressure, which is achieved by a hermetically sealed canister, acts on one side of a membrane. On the other side, the effective pressure acts and thus generates a deformation. The greater the difference between the reference pressure and air pressure, the greater the deformation. This deformation can now be measured and is a measure for the height.
One problem here is the calibration, since the air pressure also depends on the weather. However, if one calibrates before the start, one can measure a height above the ground. This is sufficient in most cases. However, there are combined sensors

which determine the absolute height roughly with GPS and use the air pressure sensor for the fine resolution.

Received signal and reception quality
In order to measure the reception signal and the receiving quality, no special sensor is required. These data are available at the receiver. In most telemetry systems, they are an integral part.

8. Assembly and initial operation

Many considerations have been made in this book for the correct positioning of the antenna, about diversity, but also for the safe reception. This chapter shows some practical examples.
A correct initial operation includes in addition to the interference suppression of the components also a range test. This is also discussed in this chapter.

8.1 Installation in the model and cable laying

Of course, the few examples in this chapter can't represent all existing models. But some standard cases are covered, however, and it is also shown how antennas and components can be installed and positioned.

Airplane
The model aircraft 'Chubby Lady' is an ARTF (almost ready to fly) model. In particular, all RC components still need to be installed. It is controlled quite classically by rudder and elevator.

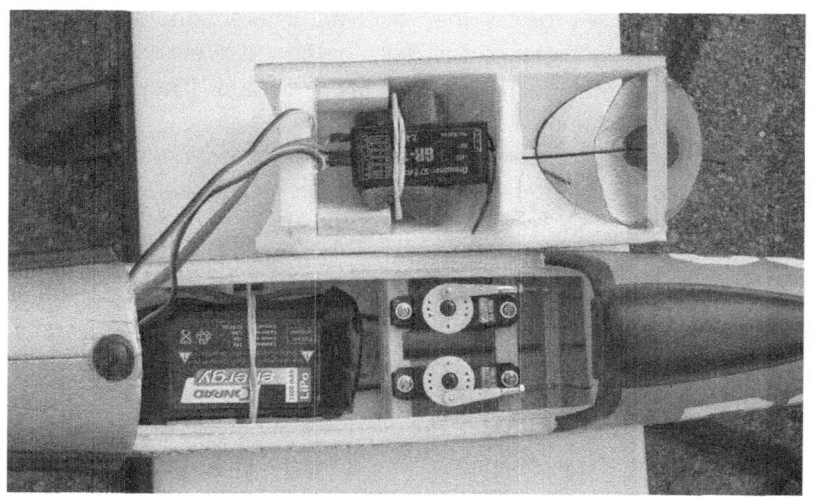

Figure 81: Assembled reception system

The brushless controller supplies the receiver with the flight battery power. A schematic illustration of the power supply with BEC was shown in Figure 3. As discussed in Section 4.5 in a mnemonic, in model aircraft diversity receivers should be used. The two antennas of the receiver are mounted at an angle of 90° to each other. This was also discussed in Section 4.5. They stick out from the model at a length of a few centimeters.

In a model aircraft which is made of a foamed material like the one in the example, one antenna can be mounted upwards. However, conductive parts worsen reception. If the fuselage is made of conductive parts, the assembly of the antenna on the bottom is useful, because the pilot often has a visual contact with the bottom of the model during flight. An operating range test should always be performed here.

Interior view

Figure 81 shows the inside view with opened hood. It should always be a principle to fix the components in the aircraft. One should take it into consideration specifically with airplanes. These can even take other positions, for example an upside-down flight.

The receiver is packed in some foam and is fixed with a rubber band. The battery is also fixed with a rubber band and with the closed hood it lies below the receiver. There should be some safety margin because of the anticipated heat emission during discharge. Behind, the two servos are mounted. They move the rudder and elevator by a linkage.

Often it is written in operating manuals that one should take great care to place current-carrying conductors the greatest possible distance from the receiver and the servos. This is well implemented in this example. At the front in the engine compartment and not visible in the figure, the brushless controller is located. So from the battery, all the power lines run directly forwards to the controller and the motor. Only the receiver power is fed back via the BEC plug. This is of course also necessary because the brushless controller also receives information on the throttle through the same cable.

The mounting of the servos is made for a long time by rubber grommets and inserted hollow brass bearings. These are usually fixed with screws into a wooden strip. Thus, the servo sits firmly in the model and at the same time is also protected against vibration.

Model helicopters

Figure 82: Model helicopter with diversity receiver

A diversity receiver is strongly recommended in model helicopters as well. Figure 82 shows a possible variant. The two antennas are mounted in two white plastic tubes at the rear. Again, the 90° orientation to each other is apparent. The last few centimeters are free like with the airplane model. Helicopters are often made of metal or carbon fiber parts. Just as the antennas should not be in the vicinity of cables, they should generally be located as far away as possible from conductive materials. This factor is taken into account in this example. As the figure shows, they are installed with a large distance to the metal parts.

For the same reason, they are also mounted to the bottom of the helicopter. Since during flight the model pilot is below the model, there is visual contact between the transmitter and receiver antennas.

Motorized paraglider

Figure 83 shows the model of a motorized paraglider. Again, the antennas are brought out to both sides from the model. Only one is visible here.

Such freely hanging-out antennas can be seen very often. They can then move in the wind during flight and there is no guarantee that they are 90° apart. If they were in plastic tubes as in Figure 82, they could be mounted to the rear, slightly angled from the fuselage as there. Again, it would make sense that the last few centimeters stick out of the tube.

Figure 83: Motorized paraglider

Cars and boat models

Model cars and model boats often do not have diversity receivers. As described in Chapter 4, the antennas are often vertically polarized. Then they point vertically upwards. Figure 84 shows that for a ship model.

Figure 84: 2.4 GHz antenna on a model boat

It can also be seen that the antenna must be mounted so that it is waterproof. The receiver is usually located very close to the feedthrough inside the housing, and water is something that electronics don't like at all. For this reason, the plastic tube is screwed and connected to a sealing ring. In addition, the antenna itself is also provided with a shrink tube so that in between no water can penetrate into the hull.

Figure 85 shows the antenna assembly in a car model. Since the chassis is not made watertight, the receiver is housed in a box protected against splashing water. In this example, the BEC cable with ferrite core and that of the steering servo lead into the housing. As with all applications in which the receiver is supplied with the BEC, the motor controller also receives its signals via this cable. This again corresponds to the illustration in Figure 3 with the difference that only one servo is used. The antenna is fed out in a rubber hose to the top.

Figure 85: Antenna on a car model

Antenna mounting inside the model?

Sometimes you can see in practice antennas that are mounted inside the hull. That may work with foam models or those with plastic covers. However, you should then always perform a test of the operating range and compare it with an externally mounted antenna. With inside installations, the cables, the servos and the drive are closer to the antenna and impact on the reception. With any model hulls made of conductive materials such as carbon fiber, the internal mounting of the antenna is strongly discouraged. The material penetration of the 2.4 GHz radio waves is simply worse than that of MHz waves. Therefore all the antennas which are described in the examples are directly visible from the outside, even if you would maybe prefer to avoid this for aesthetic reasons. In today's sometimes very large models, the reliability of reception is the most important consideration and one should therefore do everything possible to ensure that it is always optimally guaranteed. The author always mounts the antennas outside of his models.

8.2 Interference suppression of components

Ferrite core in the BEC cable
In many motor controllers, the BEC cable is already equipped as standard with a ferrite core. This is a ring of iron. The wire should be wound several times in the same direction around the core. A ferrite core is shown in Figures 11 and 85. It dampens interferences that come from the motor controller so that the receiver gets a clean voltage. With reception problems that is a possibility for better interference suppression.

Interference suppression of electric motors
Brushed DC motors are today only rarely used as drives of models because they have some disadvantages in terms of efficiency and wear compared to brushless motors. Nevertheless, they are still installed in a few models. Since the mechanical switching of current flow through the brushes causes high voltage peaks and thus disturbances over the entire frequency range, these motors must be interference suppressed. Interference suppression is usually done by three capacitors. Two are led between the two connectors and the housing. A third is located in between them. The size of the capacitors is dependent on the size of the motor. If motors for special functions are used in a model, such as retractable landing gears, cranes, water pumps or the like, they must of course be properly interference suppressed.
Today, brushless motors are used almost everywhere as drives and they do not need interference suppression. The commutation which causes the voltage peaks in the DC motor is here realized electronically. No voltage spikes will occur because of circuitry measures in the brushless controllers.

8.3 Range test

The test of the range is today clearly part of the checklist which a dutiful model pilot or captain should perform before each use of his model. Most radio controls provide test functions for this. In most solutions the transmitter power is greatly reduced, so a possible interference is already noticeable after some ten meters.

It is very important that a model can never be started in this mode, since otherwise this inevitably leads to a loss of reception after maybe 50 meters.

After switching on, some RCs go into this mode only for a short time. For other ones this mode is only possible again after a shutdown, when the transmitter has already sent at full power. In other RCs a special key combination must be pressed to perform the range test at reduced transmitter power.

The test is described in the operating manual. It is always similar: the model should be slightly elevated, perhaps standing on a carton or a wooden box. The raised position is necessary because of the Fresnel zone (reminder Section 3.3).

Now you should move away from the model, at first with the motor switched off. You must move the stick and always watch the servo movements. The maximum distance depends on the reduced transmitter power of the range test, but it is given in most manuals as approximately 30 to 50 meters. Up to this distance the servos should follow the movement.

If the model is equipped with a motor, you should perform the same again when it is turned on. Here too, the servos should follow the stick movements, up to the maximum distance. Optionally, a helper can check the servo movements in the model. Now you can also move further than up to the maximum distance, until the servos start fluttering. After switching to the normal transmission mode, everything should work again normally, even at this distance.

If the range test fails
Sometimes the servos flutter even in normal operation, or the receiver turns into the fail-safe mode, or this happens in the range test already at less than the maximum distance. Then the following things should be checked again:

- Is the receiver antenna mounted far enough away from conductive parts and current-carrying conductors?
- Are the current-carrying conductors placed sufficiently far away from the receiver?
- Where a brushed motor is being used: is it correctly interference suppressed?
- Does it help to insert a ferrite core in the BEC cable?

If these points are resolved, the radio control should work reliably.

9. Literature

Benfield, Matt: Radio-Control Car Manual: The complete guide to buying, building and maintaining. Haynes Publishing 2008. ISBN 978-1-84425-470-5

Breck, Baldwin: DIY RC Airplanes from Scratch. McGraw-Hill/TAB Electronics 2013. ISBN 978-0-07181-004-3

Cundell, John: Radio Control in Model Boats. Special Interest Model Books 2011. ISBN 978-1-85486-231-0

Frohn, Siegfried: Fernsteuerungen im Schiffsmodellbau für Ein- und Umsteiger. Verlag für Technik und Handwerk, Baden-Baden 2010. ISBN 978-3-88180-420-2

Judd, F.C.: Radio Control for Model Ships, Boats and Aircraft. Pomona Press 2011. ISBN 978-1-44741-128-4

Kainberger, Gerald: Das grosse Buch des Modellflugs. Verlag für Technik und Handwerk, Baden-Baden 2010. ISBN 978-3-88180-793-7

Kark, Klaus: Antennen und Strahlungsfelder: Elektromagnetische Wellen auf Leitungen, im Freiraum und ihre Abstrahlung. Vieweg + Teubner Verlag, 4. Auflage 2011. ISBN 978-3-83481-495-1

Kemme, Gerhard: Von Antenne zu Antenne: Notizen zu einer Theorie der Übertragung. Elektromagnetische Wellen auf Leitungen, im Freiraum und ihre Abstrahlung. Books on Demand, Norderstedt, 2. Auflage 2009. ISBN 978-3-83703-862-0

Kotting, Manfred- Dieter: Moderne Fernsteuerungen für RC-Flugmodelle: Empfänger, Servos, Zubehör. Verlag für Technik und Handwerk 2008. ISBN 978-3-88180-780-7

Meyer, Martin: Kommunikationstechnik: Konzepte der modernen Nachrichtenübertragung. Vieweg + Teubner Verlag, 4. Auflage 2012. ISBN 978-3-83481-338-1

Molisch, Andreas F.: Wireless Communications. Wiley – IEEE 2010. ISBN 978-0-47074-186-3

Passern, Ulrich: Das LiPo-Buch. Verlag für Technik und Handwerk, Baden-Baden 2012. ISBN 978-3-88180-434-9

Sauter, Martin: Grundkurs Mobile Kommunikationssysteme: UMTS, HSDPA und LTE, GSM, GPRS und Wireless LAN. Vieweg + Teubner Verlag, 4. Auflage 2011. ISBN 978-3-83481-407-4

Tietze, Ulrich / Schenk, Christoph / Gamm, Eberhard: Halbleiter- Schaltungstechnik. Springer Verlag, 14. Auflage 2012. ISBN 978-3-64231-025-6

Wong, K. Daniel: Fundamentals of Wireless Communication Engineering Technologies (Information and Communication Technology Series). Wiley 2012. ISBN 978-0-47056-544-5